Food intolerance and the food industry

Related titles from Woodhead's food science, technology and nutrition list:

Food labelling (ISBN: 1 85573 496 6)
J Ralph Blanchfield

Food labelling has become a complex and controversial area. This collection draws on the experience of key experts in their field to provide food manufacturers with a framework within which to plan labelling effectively. It covers both the key legislation they need to be aware of, and the issues they need to take account of in successful labelling.

Principles and practices for the safe processing of food (ISBN: 1 85573 362 5)
H J Heinz Company

This food industry handbook is a practical, science-based guide to food safety in food processing operations. The text is organised for easy reference, illustrated with numerous schematics and includes important reference data tables.

'... readers will want to have this book, not just because it is such a comprehensive text on safe processing, but it is so full of sound advice. For food companies working on HACCP programmes, this book is a must-have.' *Food Engineering*.

Instrumentation and sensors for the food industry (ISBN: 1 85573 363 3)
E Kress-Rogers

The food industry is placing an increasing emphasis on measuring, controlling and improving food quality; selecting the correct instrumentation is central to this process. This book provides an authoritative and practical guide to the range of instrumentation and sensors available to the food industry professional. Topics covered include colour, texture and rheological measurement as well as composition and microbiological analysis.

'International experts have presented their knowledge in an applied framework to provide the most comprehensive workbook for practitioners ever written.' *Food Science and Technology Abstracts*.

Details of these books and a complete list of Woodhead's food science, technology and nutrition titles can be obtained by:

- visiting our web site at *www.woodhead-publishing.com*
- contacting Customer Services (e-mail: *sales@woodhead-publishing.com*; fax: +44 (0)1223 893694; tel: +44 (0)1223 891358 ext. 30; address: Woodhead Publishing Ltd, Abington Hall, Abington, Cambridge CB1 6AH, England)

If you would like to receive information on forthcoming titles in this area, please send your address details to: Francis Dodds (address, tel. and fax as above; e-mail: *francisd@woodhead-publishing.com*). Please confirm which subject areas you are interested in.

Food intolerance and the food industry

Edited by
Taraneh Dean

CRC Press
Boca Raton Boston New York Washington, DC

WOODHEAD PUBLISHING LIMITED
Cambridge England

Published by Woodhead Publishing Limited
Abington Hall, Abington
Cambridge CB1 6AH
England

Published in North and South America by CRC Press LLC
2000 Corporate Blvd, NW
Boca Raton FL 33431
USA

First published 2000, Woodhead Publishing Limited and CRC Press LLC

British Library Cataloguing in Publication Data
A catalogue record for this book is available from the British Library.

Library of Congress Cataloging-in-Publication Data
A catalog record for this book is available from the Library of Congress.

Woodhead Publishing Limited ISBN 1 85573 497 4
CRC Press ISBN 0 8493 0853 4
CRC Press order number: WP0853

Cover design by The ColourStudio
Project managed by Macfarlane Production Services, Markyate, Hertfordshire
Typeset by MHL Typesetting Limited, Coventry, Warwickshire
Printed by T J International, Padstow, Cornwall, England

Contents

Contributors

Chapter 1

Dr Taraneh Dean
The David Hyde Asthma & Allergy
Research Centre
St Mary's Hospital
Newport
Isle of Wight PO30 5TG

Tel: +44 (0)1983 534187
Fax: +44 (0)1983 534907
E-mail: dean@port.ac.uk
tara.dean@iowht.swest.nhs.uk

Chapter 3

Dr Belinda Bateman
The David Hyde Asthma & Allergy
Research Centre
St Mary's Hospital
Newport
Isle of Wight PO30 5TG

Tel: +44 (0)1983 534187
Fax: +44 (0)1983 534907

Chapter 4

Dr Jonathan Hourihane
Institute of Child Health
30 Guildford Street
London WC1N 1EH

Tel: +44 (0)20 7242 9789
Fax: +44 (0)20 7813 8494
E-mail: J.Hourihane@ich.ucl.ac.uk

Chapter 5

Dr Hasan Arshad
The David Hyde Asthma & Allergy
Research Centre
St Mary's Hospital
Newport
Isle of Wight PO30 5TG

Tel: +44 (0)1983 534373
Fax: +44 (0)1983 822928
E-mail: sha@soton.ac.uk

Chapter 6

Ms Fiona Angus and J. Smith
Leatherhead Food Research Association
Randalls Road
Leatherhead
Surrey KT22 7RY

Tel: +44 (0)1372 376761
Fax: +44 (0)1372 386228
E-mail: fangus@lfra.co.uk

Chapter 7

Dr Sally A Kilburn
School of Postgraduate Medicine

Gloucester House
Queen Alexander Hospital
Cosham PO6 3LY

Tel: +44 (0)23 9228 6236
Fax: +44 (0)23 9228 6227
E-mail: sally.kilburn@port.ac.uk

Chapter 8

Ms Johanna Hignett
Nestle UK Limited
St George's House
Croydon
Surrey CR9 1NR

Tel: +44 (0)20 667 5532
Fax: +44 (0)20 8667 6061
E-mail: johanna.hignett@nestlegb.nestle.com

Chapter 9

Mr David Reading
The Anaphylaxis Campaign
PO Box 149
Fleet
Hampshire GU13 9XU

Tel: +44 (0)1252 542029
Fax: +44 (0)1252 377140

Chapter 10

Drs Abbas Khakoo, Graham Roberts and Gideon Lack
Department of Paediatric Allergy and Clinical Immunology
St Mary's Hospital
Praed Street
London W2 1NY

Tel: +44 (0)20 7886 6384
Fax: +44 (0)20 7886 1129

Abbreviations

AAA	Action Against Allergy
ACE	angiotensin converting enzyme
ADSA	Association for Dietitians in South Africa
ALBA	Dutch Food Intolerance Databank
BAF	British Allergy Foundation
BCA	bicinhoninic acid
BDA	British Dietetic Association
BHA	butylated hydroxyanisole
BHT	butylated hydroxytoluene
BRC	British Retail Consortium
CCP	critical control point
CMA	cow's milk allergy
DBPCFC	double-blind placebo-controlled food challenge
DNA	deoxyribonucleic acid
EAACI	European Academy of Allergology and Clinical Immunology
EC	European Commission
EFID	European Food Intolerance Databanks
EFLA	European Food Law Association
eHF	extensively hydrolysed formula
ELISA	enzyme-linked immunosorbent assay
EU	European Union
FARRP	Food Allergy Research and Resource Program (University of Nebraska)
FCAA	Food and Chemical Allergy Association
FCPMC	Food and Consumer Products Manufacturers of Canada
FDA	(US) Food and Drug Administration

FDF	Food and Drink Federation
FEIA	fluorescent enzymatic immunoassay
FEV1	forced expiratory volume in 1 second
FIDB	Food Intolerance Databank (South Africa)
FLAG	Food Labelling Agenda
FLEP	Food Law Enforcement Practitioners
FSA	Food Safety Act
GMP	Good Manufacturing Practice
GP	general practitioner
GPSR	General Product Safety Regulations
HACCP	hazard analysis and critical control points
HF	hydrolysed formula
HLA	human leucocyte antigen
Ig	immunoglobulin
JECFA	Joint Expert Committee on Food Additives
LIVO	National Information Centre for Food Hypersensitivity (Netherlands)
MAFF	Ministry of Agriculture, Fisheries and Food
ME	myalgic encephalomyelitis
NHS	National Health Service
OAS	oral allergy syndrome
PARNUT	food for particular nutritional use
PEFR	peak expiratory flow rate
pHF	partially hydrolysed formula
RA	Research Association
RAST	radioallergosorbent test
RCP	Royal College of Physicians
RIA	radioimmunoassay
RNA	ribonucleic acid
SPT	skin prick test
TEI	Technological Educational Institution Thessaloniki
TGF	transforming growth factor
Th	T-helper
TNO	Netherlands Organisation for Applied Scientific Research

1

Introduction

T. Dean, The David Hide Asthma and Allergy Research Centre, Isle of Wight

1.1 Background

The documentation of food intolerance goes back to 55 BC when Lucretius, a distinguished Latin poet and philosopher, wrote his poem *De Rerum Natura* (On the Nature of Things) and said 'What is food for some, may be fierce poison for others'.

Hippocrates recognised the adverse effects of milk on some individuals when he noted:

> Cheese does not harm all men alike; some can eat their fill of it without the slightest hurt. ... Others come off badly. So the constitutions of these men differ, and the difference lies in the constituent of the body which is hostile to cheese, and is roused and stirred to action under its influence. ... But if cheese were bad for the human constitution without exception, it would have hurt all.

In 1808 Robert Willan described a case where a severe allergic reaction was provoked by eating a small amount of almonds:

> These symptoms were soon followed by an oedematous swelling of the face, especially of the lips and nose, which were very hot and itchy. There was at the same time an uneasy tickling sensation in the throat, which excited a troublesome cough and a constriction of the fauces, which seemed to threaten suffocation. The tongue, likewise, became enlarged and stiff, causing slowness and faltering in the speech. Soon after going to bed an eruption took place over the whole body of spots nearly as large as a sixpence, of a dead white colour, a little elevated above the skin, like weals produced by the sting of a nettle, and intolerably itching.

There are many such anecdotes in medical history literature. What is noteworthy is that, unlike most other disciplines where scientific research starts soon after such anecdotes, in the food intolerance area there has been a large gap between the case reports and scientific investigation of the field. This has created opportunities for many people to blame food intolerance for a wide range of unexplained disorders, and for many years food intolerance was regarded to be on the fringe of scientific enquiries. The fact that for decades the diagnosis of food intolerance relied mainly on clinical history created many opportunities for individuals and groups offering all sorts of unscientific and bizarre tests for diagnosis of food intolerance. It is only fairly recently, with the introduction of double-blind placebo-controlled food challenges, that opportunities for more scientific approaches have been created and research into this area has provided us with good quality evidence.

Just as high quality research evolved in the midst of anecdotes, the terminology in this field also evolved, and terms such as food hypersensitivity, food intolerance, food allergy and adverse reactions to food are used at times interchangeably. In the next section, some of these terms are described in more detail.

1.2 Terminology

When reading different texts in this area, it becomes evident that in the medical and scientific community, there is no single global consensus on what is food allergy and what is food intolerance. For example, there are authorities who consider coeliac disease as a type of food allergy and others who regard it as a form of food intolerance. Some may not consider it as either. Indeed, it appears that it all depends on what definition one has used. The terminology which appears to have gained credibility amongst many peers is that adopted by the European Academy of Allergology and Clinical Immunology (EAACI).[1] The distinguishing feature of this terminology is that it is based on mechanisms rather than clinical symptoms. The structure of this terminology is outlined in Fig. 1.1. Broadly, adverse reactions are divided into toxic and non-toxic reactions.

1.2.1 Toxic food reactions

In principle, these are reactions which could occur in any individuals if the dose is high enough to trigger a reaction. They are usually caused by direct action of food components without involvement of immune mechanisms. Toxic compounds which trigger such reactions can occur naturally, such as from eating a puffer fish complete with its poison sac! Or they can be contaminants of food. Although such reactions are fairly distinguishable from non-toxic food reactions in terms of mechanism, one has to be careful when diagnoses are made, since some of the symptoms may be similar.

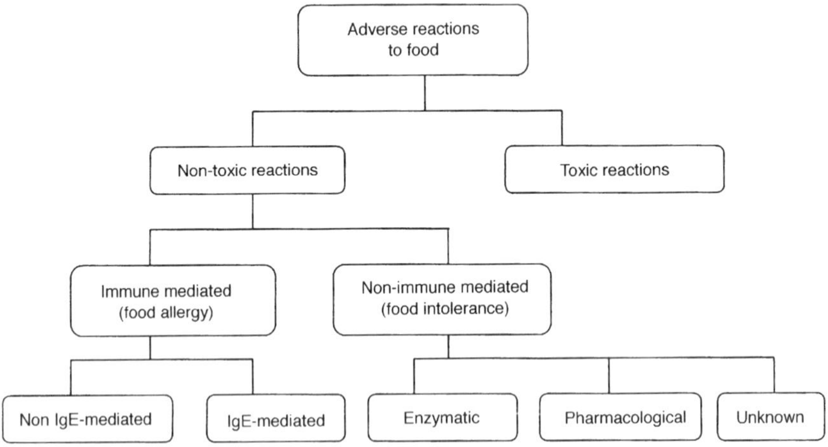

Fig. 1.1 Classification of adverse reactions to food.

1.2.2 Non-toxic food reactions

These reactions are either immune-mediated or non-immune-mediated. When the reaction is immune-mediated the term 'food allergy' is applied, and when non-immune-mediated the term 'food intolerance' is recommended. Both types of reactions are reproducible and depend on an individual's susceptibility.

Food allergy

Food allergy itself can be subdivided into two categories, IgE-mediated food allergy and non-IgE-mediated food allergy (Fig. 1.1). Immunoglobulin (Ig) E, or IgE, is the main antibody involved in induction of immediate allergic reactions. Most of the research evidence available on food allergy has been focused on IgE-mediated food allergy. Indeed, most common food allergies are mediated by IgE antibodies. The mechanism underlying IgE-mediated food allergy is fairly well established. Symptoms of this form of food allergy appear rapidly, are varied and range from anaphylaxis to skin reactions.[2]

Non-IgE-mediated food allergy is less well understood. Such allergies include reactions involving other immunoglobulin isotypes such as IgG and its subclasses, food immune complexes and cell-mediated immunity. Diagnosing this form of food allergy has been difficult and none of the above-mentioned mechanisms have been proven to be causative by double-blind, placebo-controlled food challenges (DBPCFC).

Food intolerance

Food intolerance reactions are reproducible non-immune-mediated reactions which, depending on their causality, are divided into the following types of intolerance:

- enzymatic
- pharmacological, i.e. reactions caused by either naturally derived or added chemicals that produce a pharmacological effect in the individual
- undefined food intolerance.

Lactase deficiency (usually referred to as lactose intolerance) is a good example of the enzymatic form of food intolerance.[3] It is often secondary to other conditions such as viral gastroenteritis. In very rare situations lactase deficiency can be an inborn error of metabolism.

Examples of pharmacological forms of food intolerance include reactions to vasoactive amines, such as histamine, found in many foods. The importance of these amines in provocation of symptoms is not well defined. The third category of food intolerance reactions is 'undefined'. These reactions include any reproducible adverse reaction due to an unknown mechanism. Reactions to food additives may be considered in this category.

1.2.3 Food aversion

This includes food avoidance and psychological intolerance. These reactions are not truly food dependent and are excluded from the EAACI[1] classification. Food aversion is not a reproducible reaction, and if the offending food is disguised the reaction will not take place. In reality, a large proportion of people who believe they are allergic belong to this category and continue to avoid the offending food without dietetic supervision, sometimes with adverse nutritional consequences!

1.2.4 Other terms

The term 'food sensitivity' is used mainly in North America and is defined as any adverse reaction to food. This is subdivided into 'food hypersensitivity' (immunologically mediated) and food intolerance (non-immune-mediated). Although not a popular term outside the US, it is important to be familiar with the definition of food sensitivity, in particular when reading books and articles from North America.

1.3 Mechanisms of food intolerance and food allergy

With regard to underlying mechanisms and trigger factors for food allergy and food intolerance, it is fair to say that our level of knowledge is very much in its infancy. We know, for example, that some individuals are more susceptible than others. Atopy (predisposition to allergic disease) is heritable, so could this susceptibility be due to genetic factors? What about environmental factors, in particular during infancy? What is their impact?[4]

What role do food allergens themselves play? We know generally that the most common foods implicated in food allergy and food intolerance are egg, milk, peanuts, nuts, fish and soya.[5–7] On average, an individual's gastrointestinal tract will process about 100 tonnes of food during a lifetime. Everything we eat is foreign to our body and potentially immunogenic. What is so special about some food allergens? Why do only a proportion of people have the ability to sensitise and cause an allergic reaction? What is the natural history of food allergy and food intolerance? We do not know why with some foods, such as milk, sensitivity is lost with time, while with others, such as peanut, the reaction seems to be long-lasting. What influences do our eating habits have on our allergy and intolerance profile? For example, soybean and rice allergy is more common in Japan, and fish allergy is more prevalent in Scandinavian countries. Who in the UK had heard of kiwifruit allergy 20 years ago?

These are some of the unresolved issues in the area of food allergy and intolerance. Much work is needed to answer these questions and understand the basic mechanisms involved in this area.

In the following section some of our basic understandings of the mechanisms of food allergy and food intolerance are summarised.

1.3.1 Immunological mechanisms

Type I: IgE-mediated reactions

These reactions are the most frequent, the best known and the easiest to diagnose. They occur when an individual is already sensitised. In susceptible individuals, when a food allergen is encountered for the first time, the adaptive response initiates production of IgE antibodies. IgE antibody production itself is regulated and depends upon compounds known as cytokines.

Once IgE antibodies are produced, they will bind to mast cells. This process, known as sensitisation, precedes symptoms of allergy. How early in life an individual can be sensitised has been a topic of much interest lately. Some would argue that sensitisation can take place *in utero*. The second stage following sensitisation can take place weeks or sometimes years later. This stage occurs when the individual encounters the same food allergen for the second time. The allergen will encounter the mast cells, which already possess allergen-specific antibodies on their surface. IgE antibodies will bind the allergen and this will lead to mast-cell degranulations and release of mediators such as histamine, and the characteristic features of allergic disease follow.[8] These include:

- urticaria (this is the specific term used for hives, which are red, itchy skin welts brought on by an allergic reaction)
- angioedema (this condition often co-exists with urticaria and usually happens when urticaria affects deeper tissues and swelling results; the most common sites are lips, tongue, eyelids and larynx)

- hypotension (low blood pressure)
- anaphylaxis (this is a severe reaction with a rapid onset causing circulatory collapse, hypotension and suffocation due to tracheal swelling).

Type II: Non-IgE-mediated reactions

Here the adverse reaction is the consequence of an immune response other than IgE. It could involve another class of immunoglobulins, or food immune complexes, or cell-mediated immunity. Other immunoglobulins implicated are IgG and their subclasses. IgG4 to specific foods are often detected in those with adverse reactions to food.[9,10] The problem is that these antibodies are quite common, in both healthy and diseased states, and are often detected in normal subjects.[11] Their presence in both allergic and non-allergic individuals is explained by the immune system's ability to produce IgG.[12] Knowing the amount of food we eat, it is not surprising that IgG4 antibodies have been demonstrated. Some suggest that the presence of IgG4 alone reflects dietary intake, and the presence of both IgG4 and IgE reflects sensitisation in the individual.[13] Further studies are needed to establish the contribution of IgG4 towards immune reactions to foods.

Type III: Immune complex-mediated reactions

These reactions are also referred to as Type III hypersensitivity reactions. When we eat, food proteins that are absorbed encounter specific antibodies in the circulation and form complexes.[14,15] These are known as immune complexes. It appears that immune complex formation is essentially a normal process that occurs in the course of an immune response and allows antigen clearing. These food immune complexes contain IgE, IgG and IgA antibodies. They are usually cleared very quickly by our reticulo-endothelial system. The pathology is caused when these complexes are deposited in certain tissue sites. Their pathogenic potential is determined partly by their size and partly by their concentrations. If these food antigens and immune complexes are present in very high concentrations, tissue damage can occur. This damage is done by activation of complement, releasing C5a to create a local inflammatory response and hence increasing vascular permeability, which allows fluids and cells to enter the site. Although circulating immune complexes containing food antigen have been demonstrated in patients with food allergy suffering from asthma and eczema, there is no definitive evidence that either IgG or IgE food immune complexes cause the disease.[16]

Type IV: Cell-mediated food allergy

These reactions are sometimes referred to as Type IV or delayed-type hypersensitivity reactions. They are mediated by inflammatory T cells. There are reports of cell-mediated immune responses to food antigens in cow's milk allergy. There is also good evidence to suggest that coeliac disease may be provoked by a cell-mediated food allergy to gliadin, a constituent part of gluten.[17] The characteristic hyperplastic villous atrophy in coeliac disease is secondary to an abnormal T cell-mediated response to gliadin. Although T cells have been shown to be involved in coeliac disease,[18] there is little evidence to suggest that cell-mediated food allergy to gliadin is the original cause of coeliac disease.

It is important to consider that these immune mechanisms are not mutually exclusive and more than one mechanism may very well operate at any one time.

1.3.2 Non-immunological mechanisms

Enzymatic mechanisms

Enzymatic food intolerance is due to an enzyme defect which could result from an inborn error of metabolism or could be secondary to a number of disorders. The most common food intolerances in this category are disaccharide deficiencies, galactosemia and phenylketonuria. Amongst disaccharide deficiencies, lactose intolerance is the most common. Lactose deficiency can be congenital, persisting in the neonatal period, or can be acquired where it presents later.[19] These deficiencies are genetically based and not due to an environmental factor. Secondary lactase deficiency often occurs following an episode of gastroenteritis.

Galactosemia is also a form of carbohydrate deficiency.[20] Here the genetic imbalance is expressed as a deficiency of either galactokinase, galactose 1-phosphate uridyl-transferase or uridine diphosphate galactose 4-epimerase. These enzymes are responsible for converting galactose to glucose. So in individuals with this deficiency this process fails. The clinical manifestations are toxicity syndromes when exposed to galactose. These include failure to thrive, vomiting and liver disease.

Phenylketonuria is caused by a gene mutation which suppresses the activity of phenylalanine hydroxylase enzyme.[21] Approximately 4–5% of amino acids in all food protein are phenylalanine. Restriction of phenylalanine intake to 0.4 mmol/l throughout life ensures almost normal physical and mental development.

Pharmacological mechanisms

Many foods contain pharmacologically active components. A pharmacological food intolerance is usually evident soon after eating the food responsible. The amount of food ingested to elicit a reaction varies from person to person and may even vary in the same individual over time. The pharmacological components can either initiate a reaction directly themselves or activate the host's mediator system indirectly and hence induce a reaction.

The largest class of substances that are found in many foods responsible for inducing pharmacological food intolerance are vasoactive amines. Other substances involved are methylxanthines, capsaicin and ethanol. Vasoactive amines include histamine, tyramine, tryptamine and serotonin. Foods such as tuna, cheese (in particular Parmesan and Roquefort), yeast extracts such as Marmite, and red wine such as Burgundy and Chianti are rich in these amines. Ingestion of large amounts of these foods can be followed by toxic symptoms.

The best example is scombroid poisoning, due to an excessive amount of histamine in some species of fish such as tuna and mackerel.[22] Because histamine is a mediator released from mast cells in food allergy, sometimes pharmacological food intolerance due to histamine is confused with allergic-type reactions. In scombroid poisoning, when fish is inadequately refrigerated, marine bacteria convert the amino acid histidine to histamine. This will generate a histamine concentration greater than the body's normal capacity to metabolise and hence the individual will suffer from the full spectrum of histamine effects, including flushes, vomiting and diarrhoea.

Some foods that do not have a high histamine content themselves contain compounds that can indirectly induce degranulation of mast cells and histamine release. These foods include chocolate, ethanol, tomatoes and crustaceans.

Vasoactive amines such as tyramine and serotonin are found mainly in fermented food.

1.4 Mechanisms of oral tolerance

Oral tolerance is very much the norm. The reason why we are not all allergic and intolerant when we eat food is due to basic mechanisms that function in the development of our tolerance. Food intolerance and food allergy is in fact a failure of oral tolerance. The existence of oral tolerance has been known for a long time, but its mechanisms are still not fully understood. A number of experimental models have been used to demonstrate this phenomenon. One such example is the oral tolerance to ovalbumin in mice. This was induced by a single administration of ovalbumin and a demonstration of suppression of cell-mediated immunity.[23]

T-helper cells are differentiated into two subsets, known as Th1 cells and Th2 cells. Th1 cells produce cytokines such as gamma-interferon and induce macrophage activation. In the absence of gamma-interferon, the antigen-presenting cells express another cytokine, IL-10, and induce Th2 cells to produce IL-4 cytokines. The latter cytokines will instruct naïve B-cells to produce IgE. The balance between gamma-interferon and IL-4 at the time of the immune reaction will govern the immune outcome. High interferon/IL-4 production facilitates the induction of a Th1-type immune response, whereas high IL-4 production induces a Th2 pathway. In oral tolerance, it is suggested that T-helper cells known as Th3 type are involved. These cells, which produce TGF-β1, may be responsible for oral tolerance, since TGF-β1 downregulate inflammatory cytokines and promote IgA production.[24]

1.4.1 Factors contributing to development of tolerance

Genetic background

There are a number of factors which influence development of tolerance. Genetic background is an important factor. Atopy, defined as the genetic tendency to respond with IgE to exogenous proteins, is strongly associated with allergic symptoms. The risk of developing allergic disorders increases in children born to families with atopy. It is, however, unclear whether genetic differences in antigen clearance are associated with the capacity to induce oral tolerance.

Dose of antigen

Most of our knowledge in this area has originated from animal studies. Dose–response studies with rodent models have shown that the low-IgE responder phenotype develop specific tolerance in response to inhalation of nanograms of antigen, whereas high-IgE responders require much higher doses for tolerance.[25] Ultimately, it may be that a person's sensitivity ot tolerance is governed by their genetic background.

Time of exposure

What happens when we encounter an antigen for the very first time has an important impact on what the outcome may be, i.e. whether we are sensitised or tolerised. We encounter most antigens early in life, during infancy, and as it happens this is the period when tolerance development is impaired. In fact, this delayed post-natal maturation of tolerance has been suggested as the reason for the increased frequency of allergic symptoms in infancy.[26] So it appears that there is an immunological vulnerable period, perhaps due to the inability of the immature immune system to induce tolerance. Clearly, this is an exciting area and only further research will elucidate the current ambiguity.

1.5 Summary

This chapter has aimed to give some background to the history of food intolerance and food allergy. Hopefully it has shown that food allergy and intolerance is a condition that has existed for centuries, although it may not have been labelled as such. This was followed by a section on terminology from which the reader will recognise that the debate still continues. Although the recommendations of the European Academy of Allergology and Clinical Immunology (EAACI) are outlined in the chapter, it is likely that readers will come across other terms or indeed, more commonly, will find these terms being used in different contexts. In time, no doubt, a worldwide terminology may evolve, but for the time being clarification and definition have to be offered in the literature.

The sections which followed were dedicated solely to mechanisms of food intolerance and allergy and food tolerance. It is essential to have a grasp of the current state of evidence in this area in order to appreciate some of the key issues concerning diagnosis, symptoms, etc. Of course, it is also evident from this discussion that the medical and scientific community are nowhere near unfolding the whole picture, and our understanding of the mechanism of food allergy and intolerance is still very much in its infancy. Systematic investigation in these areas is relatively new and confined mainly to the last few decades. Our understanding can only increase with time.

1.6 References

1 BRUIJNZEEL-KOOMEN C, ORTOLANI C, AAS K, BINDSLEV-JENSEN C, BJORKSTEN B, MONERET-VAUTRIN C and WUTHRICH B, 'Adverse reactions to food, *Allergy*, 1995: **50** 623–35.

2 SAMPSON H A, 'IgE mediated food intolerance', *Allergy Clin Immunol*, 1988 **81** 495–504.

3 GRAY G M, 'Intestinal disaccharidase deficiencies and glucose–galactose malabsorption'. In Stanbury J B, Wyngaarden J B, Fredrickson D S, Goldstein J L and Brown M S (eds), *The Metabolic Basis of Inherited Disease,* fifth edition, 1729–42, McGraw-Hill, New York, 1983.

4 DEAN T P, 'Factors predicting food allergy', *Environmental Toxicology and Pharmacology,* 1997 **4** 85–9.

5 SAMPSON H A, 'Food allergy', *J Allergy Clin Immunol,* 1989 **84** 1061–7.

6 MAY C D, 'Objective clinical and laboratory studies of immediate hypersensitivity reactions to foods in asthmatic children', *J Allergy Clin Immunol,* 1976 **58** 500–15.

7 DEAN T P, 'Prevalence of allergic disorders in early childhood', *Pediatr Allergy Immunol*, 1997 **8** 27–31.

8 METCALFE D D, SAMPSON H A, and SIMON R A (eds), *Food Allergy: Adverse Reactions to Foods and Food Additives*, second edition, Blackwell Science, 1996.

9 PAGANELLI R, QUINTI I, D'OFFIZI P, PAPETTI C, CARINI C and AIUTI F, 'Immune complexes in food allergy: a critical reappraisal', *Ann Allergy*, 1987 **59**(II) 157–61.

10 VIJAY H M, PERELMUTTER L and BERNSTEIN J L, 'Possible role of IgG4 in discordant correlation between intracutaneous skin tests and RAST', *Int Arch Allergy Appl Immunol*, 1978 **56** 517–22.

11 MERRETT J, BURR M L, and MERRETT T G, 'A community survey of IgG4 antibody levels', *Clin Allergy*, 1983 **13** 397–407.

12 HALPERN G M, 'IgG4 as a blocking antibody', *Immunol Allergy Pract*, 1989 **11** 11–15.

13 SCHWARTZ R H, KEEFE M W and HARRIS N, 'Specific IgG4 and IgE in children with a history of immediate allergic reactions to cow's milk', *J Allergy Clin Immunol*, 1989 **83** 240.

14 DELIRE M, CAMBIASO C L and MASSON P L, 'Circulating immune complexes in infants fed on cow's milk', *Nature*, 1978 **272** 632.

15 PAGANELLI R, LEVINSKI R J, BROSTOFF J and WRATH D K, 'Immune complexes containing food proteins in normal and atopic subjects after oral challenge and effect of sodium cromoglycate on antigen absorption',

Lancet, 1979 **1** 1270–2.
16 CARINI C, 'IgE-immune complexes in food allergy: significance, pathogenicity and clinical considerations', *Clin Allergy*, 1987 **17** 485–97.
17 MARSG M N, 'Gluten, major histocompatibility complex, and the small intestine: a molecular and immunobiological approach to the spectrum of gluten-sensitivity ("celiac sprue"). *Gastroenterology*, 1992 **102** 283–304.
18 FERGUSON A, ARRANZ E and O'MAHONY S, 'Spectrum of expression of intestinal cellular immunity: proposal for a change in diagnostic criteria of celiac disease', *Ann Allergy*, 1993 **104** 1263–72.
19 MONTGOMERY R K, BULLER H A, RINGS E H H and GRAND R J, 'Lactose intolerance and the genetic regulation of intestinal lactase-phlorizin hydrolase', *FASEB J*, 1991 **5** 2824–32.
20 SEGAL S, 'Disorders of galactose metabolism'. In Scriver C R, Beaudet A L, Sly W S and Valle D (eds), *The Metabolic Basis of Inherited Disease,* sixth edition, 453–80, McGraw Hill, New York, 1989.
21 MRC, 'Working party on phenylketonuria', *Arch Dis Child,* 1993 **68** 426–7.
22 MORROW J D, MARGOLIES G R, ROWLAND B S and ROBERTS L J, 'Evidence that histamine is the causative toxin in scombroid poisoning', *N Engl J Med*, 1991 **324** 716–20.
23 BRUCE M G and FERGUSON A, 'Oral tolerance to ovalbumin in mice: Studies of chemically modified and ''biologically filtered' antigen'', *Immunology,* 1986 **57** 627–30.
24 KIM P H and KAGNOFF M F, 'Transforming growth factor $\beta 1$ increase IgA isotype switching at the clonal level', *J Immunol*, 1990 **145** 3733.
25 HOLT P G, VINES J and BRITTEN D, 'Sublingual allergen administration. I. Selective suppression of IgE production in rats by high allergen doses', *Clin Allergy* 1988 **18** 229.
26 SOOTHILL J F, STOKES C R, TURNER M W, NORMANN A P and TAYLOR B, 'Predisposing factors and the development of reaginic allergy in infancy', *Clin Allergy*, 1976 **6** 305.

2

The legal context: due diligence

M. Law, Law Laboratories Limited, Birmingham

2.1 Introduction: the law and food intolerance

The results of a food intolerance condition can vary from mild discomfort through severe pain to tragedy. How can the law help to regulate this situation and protect the consumer whilst providing a framework in which business can operate? How successful is it in achieving this objective?

Throughout history laws have existed to protect the consumer against the adulteration of food, whether deliberate or accidental. Watering down of milk and the contamination of food with heavy metals have long been the subject of investigation and prosecution. How does this translate into modern life and the problems of food intolerance? The first point to be clear about is that with a few minor exceptions, the law does not specifically recognise or refer to the problem of food intolerance and allergic reactions. It is therefore necessary to examine the legal provisions that do exist in order to see where they can be of help to the sufferer and provide protection against inadvertent consumption of a food which may give rise to a reaction.

Since 1991 the mainstay of food legislation in the UK has been the Food Safety Act 1990 which represented a significant step forward in the enforcement of safe food legislation. The Act came into force in response to intense media pressure following a number of food scares such as *Salmonella* in eggs, *Listeria* in pâté and soft cheeses, and unfit kangaroo meat which was reported to have found its way into pies and burgers. The Food Safety Act is the umbrella legislation for much of the subsidiary food legislation in the UK, including the Food Labelling Regulations 1996 which implement EC Directive 79/112 on the labelling of foodstuffs. The primary function of the labelling legislation is to inform about the true nature of foods and to provide details of the ingredients which they contain. More help for food intolerance sufferers can, however, be found in the General Product Safety Regulations 1994. These Regulations require producers to place on the market only safe products. In a modern society where food intolerance is recognised as a problem affecting a significant sector of the population, it is only right that a product which places part of that population in jeopardy is recognised as being an unsafe product which should be subject to certain controls.

With the ever widening range of products on the market and the demand for greater innovation, it would seem that all of the cards are stacked against the food manufacturer. Some kind of balance is required to ensure that the food manufacturer who uses his best endeavours to produce a safe, properly labelled product has some protection from the law. The due diligence defence is a feature of modern consumer law which provides for acquittal where certain steps have been taken to avoid the commission of the offence.

All of the law mentioned so far provides a legal framework to protect the consumer and to seek sanctions on behalf of society when all is not as it should be. This is of little help for the consumer who has suffered pain and anguish from the carelessness or negligence of others. Damages to provide recompense for a loss are the province of the civil courts. Since the mid-1980s European law has provided a straightforward remedy in the shape of product liability law which provides a clear path of liability on the part of the manufacturer, or in some cases the supplier, where damage has been caused by a defect in the goods.

2.2 The legal background: the Food Safety Act 1990

Since the days of Magna Carta there have been controls over the sale of food in one form or another. The right to 'one measure throughout the land' was an early example of this. Since that time there have been legal controls to prevent the adulteration of food. Flour and milk were early examples, to prevent the addition of chalk to flour (later required by law to boost the calcium content), and to prohibit the addition of water to milk. Since that time the technology of food and the structure of our society has become infinitely more complex. As a consequence issues which once constituted clear breaches of the law are now less easy to discern. We are now in a situation of needing to exercise judgement in order to decide whether or not a situation which may be prejudicial to some will actually give rise to an offence, or whether some other course of action may be open to the consumer. In examining the issue of food intolerance, we need to ask ourselves whether food which may be perfectly wholesome for the majority of the population may give rise to the commission of an offence under criminal legislation when it has adverse effects upon others.

The UK Food Safety Act 1990 creates certain offences. Included among these are those of:

- rendering food injurious to health;

- selling food which fails to comply with the food safety requirement because it has been rendered injurious to health, or is unfit for human consumption;
- selling food that is so contaminated that it is not reasonable to expect it to be used for human consumption;
- selling food which is not of the nature, substance and quality demanded by the purchaser; and
- giving a misleading label with food.

These provisions are intended to protect the consumer from deliberate and accidental chemical and microbiological contamination, from foreign bodies in food, from food of unacceptable quality, and from being misled. The initial problem is that a food which is perfectly safe for the majority of the population can present problems for others. This raises the question of whether or not such food is 'unsafe'. Where does allergenicity fit into this? Put simply, it does not: the offences are intended to capture mainstream contamination and abuse. Whilst it would be possible to fit certain specific situations into the law, consumers who have a specific problem need to look at avoidance rather than rely on the law to eliminate certain foods or ingredients from their diet.

2.3 The legal background: labelling

The practical protection which individuals with food allergy and intolerance can expect from the law is information rather than elimination. To this end, comprehensive food labelling requirements have developed. Throughout the European Union these requirements are largely harmonised and stem from EC Directive 79/112 on the labelling and presentation of foodstuffs. The provisions are enacted within the UK as the 1996 Food Labelling Regulations. Although it originated two decades ago, the Directive and its enactments in EU Member States have been progressively updated over the years. The legislation requires that all foods are labelled with either a legally provided name or a customary name which is well understood by purchasers in the place of purchase, or a true name which accurately describes the food. A list of ingredients is required for *most* foods which details what they contain, including any additives. Notice the emphasis on the word 'most': as with many requirements, there are exceptions, and these exceptions may mask the presence of ingredients which may result in unpleasant, dangerous or fatal consequences for a minority of consumers. The first exemption is that for compound ingredients. Any ingredient which itself consists of two or more ingredients, and does not constitute more than 25% of the finished product, may be labelled by its name alone, without the requirement for all of its constituent ingredients to be labelled. This is subject to the proviso that any additives which are functional in the finished product must be disclosed. By way of an example, consider a ready meal which lists amongst its ingredients:

Fish Sauce (contains Preservative: Sodium Benzoate)

This fish sauce may contain many ingredients which might include shellfish capable of causing an allergic reaction in susceptible people, yet the labelling meets the requirements of the law. Also, the stipulation for the labelling of functional additives may itself give rise to problems. A garlic purée used in garlic bread may have contained sulphur dioxide as a preservative, but because this preserving effect is no longer required in the finished product, possibly because it is frozen, there is no need to label its presence. This may present a hidden problem for asthmatics. Other exemptions may be realised through the provisions which permit the use of generic names for certain ingredients or because the ingredient is a food which itself is not required to be labelled with an ingredients list. Certain foods such as chocolate currently fall outside the requirements of food labelling law and are subject to the specific requirements of their own legislation. Typically, this may not require the food to be marked with a full list of ingredients, but only to disclose the presence of certain ingredients.

The European Commission has already recognised the need for transparent labelling by requiring that any starches or modified starches which contain wheat gluten are declared as such. Thus a potato starch can be described as starch or modified starch, but wheat starch must be described as wheat starch or modified wheat starch. Many retailers in the UK, together with some manufacturers, are anticipating the needs of their customers by voluntarily providing information about potential allergens in their product. This may be by providing a full breakdown of compound ingredients, by highlighting the presence of potential allergens or by making specific claims such as 'gluten free' on the product. Inevitably the law is slow to respond to the needs of consumers, and this kind of initiative can provide a useful means of communicating helpful information.

Although these labelling requirements originate in Directive 79/112/EEC on the labelling and presentation of foodstuffs, there are differences in the way that the legislation is both enacted and enforced between one Member State and another. These differences are largely manifested as extra requirements of a domestic nature, although in recent years greater efforts have been made to harmonise the requirements. In the UK, for example, all the requirements relating to claims and misleading descriptions are purely domestic in nature and will not be found in the Directive.

In 1968 when the Trade Descriptions Act came into force in the UK it was hailed as a consumers' charter. Indeed, its impact went far beyond its own provisions, for it heralded an age of consumer awareness and spurned the creation of civil law advice centres to deal with the influx of complaints and enquiries which did not fall within its scope. Despite being over 30 years old, the Act does not seem to be suffering from mid-life crisis and is still a much-used weapon in the enforcement armoury. It has relevance in relation to food allergenicity, as it provides similar provisions to the misleading label requirements of the Food Safety Act. Thus a label which proclaims that the food inside is nut free or suitable for coeliacs when through a deliberate act or carelessness that is not the case, may also constitute an offence under the 1968 Act. Many local authorities adopt the approach of instituting legal proceedings under both Acts in order to widen their chance of success.

2.4 The legal background: the control of food manufacture

In addition to the EC controls over the labelling of food, there are measures to control its manufacture. Directive 89/397/EEC on the Official Control of Foodstuffs deals with the manufacture of food and provides for it to be controlled at the point of manufacture. The main purpose of the legislation is to provide powers for enforcement officers to enter food production premises, inspect the operation and examine recipes in order to ensure that relevant legal provisions are being complied with. The Directive also requires each Member State to take the responsibility for food originating in its territory irrespective of its ultimate destination. As with most EC law, enforcement is a matter for individual Member States and reflects their patterns of government. In the UK enforcement is the responsibility of local authority Trading Standards and Environmental Health Departments, whereas in France it is enforced nationally by the Services de la Concurrence, de la Consommation et de la Répression des Fraudes. One of the problems highlighted has been the variation in the pattern of enforcement amongst the Member States. In an attempt to rectify this, the European Commission has been running the Karolous Programme for a number of years. This initiative allows enforcement officers to spend time with their colleagues in other Member States and observe their operating methods with a view to achieving greater uniformity of enforcement. At officer level, bodies such as FLEP (Food Law Enforcement Practitioners) provide a forum for the exchange of views on food law enforcement. Even at a local level, unofficial visits have been arranged between French and British officials in order to foster a better understanding of each other's problems and working practices. All of these moves are to be encouraged, as one of the greatest enemies of free competition and a true single market is a lack of uniformity in enforcement.

2.5 General product safety

Among the lesser known pieces of EC legislation is Directive 92/59/EEC on general product safety, implemented in the UK by the General Product Safety Regulations 1994. When the Directive was under discussion in Brussels in the early 1990s it was felt in the UK that it was an unnecessary measure. At that time the Food Safety Act was just coming into force, and at first sight it appeared that all of the provisions of the European Directive were already in place in existing Food Safety and Product Safety legislation. However, the General Product Safety Regulations became reality, and it soon became apparent that there were situations in which new offences would be created and therefore new protection would be available in certain situations. Despite the existence of the Food Safety Act, the General Product Safety Regulations do apply to food. The legislation places a duty upon manufacturers and sellers of goods supplied to consumers to place on the market only 'safe products'. A 'safe product' is defined as any product which under normal or foreseeable conditions of use, including duration, does not present any risk or only the minimum risks compatible with the product's use considered as acceptable and consistent with a high level of protection for the safety and health of persons, taking into account in particular:

- the characteristics of the product, including its composition, packaging, instructions for assembly and maintenance;
- the effect on other products, where it is reasonably foreseeable that it will be used with other products;
- the presentation of the product, the labelling, any instructions for its use and disposal and any other information;
- the categories of consumers at serious risk when using the product, in particular children.

The latter two points are particularly relevant in relation to serious allergy situations, because the legislation brings in the concept of selective risk and recognises that adequate information is a relevant factor in deciding whether or not a product is safe. Although at first glance it would seem difficult to describe a peanut butter cookie as dangerous, closer examination of the provisions of the legislation will show that in deciding whether or not goods are dangerous, regard must be paid to any warnings or information given with the goods.

Where specific EC law on the safety of goods exists (such as the Directive on toy safety), the Directive on general product safety will not override it. The Food Safety Act, however, is UK domestic law which is supplemented by the General Product Safety provisions.

A food product containing nut ingredients, which are not obvious from its name, can be sold in a catering establishment with no information other than its legal name, yet still meet the provisions of food labelling law despite the fact that it may pose a serious threat to a vulnerable sector of the population. However, if the requirements of General Product Safety legislation are taken into account, we have a product with the potential to cause harm. That danger can, however, be mitigated by providing adequate warnings. The use of prominent notices such as 'some of our products may contain nuts or nut traces – please ask staff for details', can be used to make the customer aware of the possible presence of nut ingredients in the absence of full labelling. To date, enforcement authorities have not made wide use of these provisions, but at least one large local authority in the UK has recognised the potential and has referred to the legislation in newspaper publicity aimed at achieving greater awareness of the problem amongst caterers. Again, it should be stressed that this is EC legislation which will apply throughout the European Union.

2.6 Civil remedies: the Consumer Protection Act

All of the legislation outlined so far is criminal legislation, that is to say it protects society in general. In addition to criminal law, civil provisions protect the individual by providing a financial remedy in the form of damages where death, injury, loss or damage has resulted from a faulty product. Although such a remedy has existed for many years, it required the plaintiff to prove that the manufacturer was negligent in the production of the food. Negligence has always been difficult for the ordinary citizen to establish without the powers to inspect the production facility or to see production records, assuming that they even existed. This inequality was recognised by the European Commission in the mid-1980s when the Product Liability Directive was established. Enacted in the UK by the Consumer Protection Act 1987, this piece of legislation provided a great step forward by eliminating the need to prove negligence. The prerequisites for a successful action were to be able to prove that the damage was caused by a defect in the goods. The definition of damage includes death or injury, thus bringing unsafe food within its remit. One of the first actions within the UK was brought by a person who had suffered botulism as a result of eating a hazelnut yogurt which had been prepared with contaminated hazelnut purée. The legislation also removes the need for there to be a direct contractual relationship between the two parties involved. Previously, the buyer would have had to take action against the seller, but now the person who suffered harm can take action directly against the manufacturer, despite there being no contractual link between them.

2.7 Due diligence

When the Food Safety Act 1990 came into force, the concept of 'due diligence' became a major talking point within the food industry. The concept was not, however, new to consumer law, having been available in the Weights and Measures Act 1963 and the Trade Descriptions Act 1968 as well as other consumer legislation. Due diligence is a protection available to potential defendants under the provisions of the Food Safety Act and the General Product Safety Regulations. It acts as a balance to the principle of strict liability which forms the basic tenet of consumer law. Strict liability means that the defendant is guilty whether or not he intended to commit the offence. Thus a food manufacturer who, for example, produces a product which by accident contains a piece of fibre from a conveyor belt will be guilty of an offence under the Food Safety Act irrespective of the fact that he was unaware that it had happened. This clearly represents an onerous burden for the manufacturer, but he can be acquitted if he is able to demonstrate that he has 'taken all reasonable precautions and exercised all due diligence to avoid the commission of the offence'. 'All reasonable precautions' means that a system of controls was in place, and 'all due diligence' means that it can be demonstrated that the system worked. The key words here are 'all' and 'reasonable'. It is necessary to show that *all* reasonable precautions were taken, not just some. It is also only necessary to take those precautions which are *reasonable*. The test of reasonableness is related to the size and nature of the business and also the risk which the precautions are designed to avoid. Risks which involve consumer safety are likely to be regarded by the Courts as carrying a higher priority than those involving possible financial loss. Legislation itself gives little clue to the detail of what is required in order to establish the defence, but the food industry can give thanks to a veritable army of used car dealers and importers of toys and novelties who, over the years, have attempted to use the defence in order to avoid conviction and have had their attempts scrutinised in minute detail by the Courts of Appeal. Some of our best known High Street retailers have also taken part in this process of shaping due diligence law. The learned judges in these cases have provided a number of decisions on individual points of issue which can be collated to provide a clearer insight into the standards to be met.

The decisions in these cases can be summarised into a series of key requirements as follows:

- The system must be under the control of the 'directing mind' of the business. Its operation can be delegated to senior managers but control must remain with the directors or owners of the business.

There needs to be practical demonstration of the control. Board meetings should include food safety issues as an agenda item. Issues should be discussed and minuted with a clear plan of action. Minutes of later meetings should demonstrate how the matter was resolved.

- The system must exist and be written down. It must be shown to work – a 'paper' system which looks impressive but fails to deliver practical results will not suffice.

There should be written procedures to control activities which can affect product safety and legality. Sufficient staff should be available to allow the system to work as intended.

- The system should be appropriate to the size of the business and the risks posed by its products.

'Off the shelf' systems will not do; the scope of the system will be dependent upon many complex factors.

- Responsibilities of staff should be clearly specified in job descriptions, and training should be given to ensure that staff are able to carry out those responsibilities effectively.

In order to show that the system works effectively, it is necessary to show that the staff have been trained both in the skills necessary to carry out their work and in the system itself.

- It must be proactive as well as reactive and should anticipate problems which are common to the business.

The system should recognise that things will not always go as planned within a business. There should be provision to deal with out-of-specification product and serious failures which will require the recall of a product.

- Records must be maintained to demonstrate that the system works as intended.

Records should relate to critical areas which can affect product safety and legality.

- The system must include the control of suppliers.

Control can be exercised through raw material specifications, supplier audits, questionnaires, certificates of compliance or analysis, or an appropriate combination of these. Reliance upon the reputation of the supplier will not suffice.

- Product testing should be a feature of the system where it is necessary to demonstrate that particular requirements have been complied with.

The level of product testing should be appropriate to the risk. The greater the level of potential risk, the greater the level and frequency of sampling.

- Complaints should be recorded and analysed in order to detect trends which should then be acted upon.

Complaints should not be regarded as a source of annoyance but as a barometer of how well the business is performing in terms of meeting both legal requirements and customer satisfaction. Problems should be carefully examined to look for their cause and a programme of improvement implemented in order to eliminate or reduce the problem.

- The system should be reviewed regularly to ensure that it remains relevant to the needs of the business.

No business is static; the laws relating to that business will change as will the nature of the business, its range of products, technology and the expectations of its customers. Systems will therefore need to be modified and updated in order to keep pace with this change.

The full requirements of the defence need to be met by food manufacturers as well as importers into the UK who will be treated as bearing the legal responsibility for the products which they import. Although the expectations of 'due diligence' are the same for imported goods as they are for goods manufactured in the UK, it is plainly not possible to discharge the responsibilities in the same way. This will require the use of auditing, holding detailed product specifications and regular testing in order to demonstrate the appropriate level of control. The greater the level of cooperation between the exporter and the importer, the easier this will be to achieve.

'Due diligence' is a principle unique to UK law. Even though its application may not be used so directly outside the UK, adoption of its principles can have advantages. For example, the principles form a sound foundation on which to organise the controls within a food business wherever it is located. If that business is supplying into the UK, adoption of the 'due diligence' principles could contribute to the UK importer's defence and make it easier to establish. Anyone familiar with quality systems will see the obvious similarities between the requirements of the 'due diligence' defence and a well-constructed quality system.

2.8 The practical application of 'due diligence' to food allergenicity

The key to practical implementation of the 'due diligence' defence is knowing your product. Modern foods are complex and will rarely be manufactured from a handful of fresh ingredients derived from known sources. Frequently, complex flavouring compounds and other bought-in functional ingredients will be used in order to provide the specific manufacturing properties and product attributes which are necessary to make a product successful in today's competitive world. It is therefore important to hold detailed specifications for these ingredients in order to be confident that they do not contain potential allergens or, at least, that adequate warning can be given if necessary. Herein lies a problem, because many manufacturers of such ingredients will be shy about revealing the exact nature of their product. A request for a product specification may result in a polite refusal as the ingredient manufacturer plays the confidentiality card. In this situation there are a number of options available. Persistence frequently pays dividends, particularly if the reasons why the information is required are carefully explained. It may help to ask for details of what is in the product, stressing that you are not interested in relative quantities. An alternative approach would be to send a dietary intolerance questionnaire to your supplier seeking details about the presence or absence of known allergens. Should this fail, you should question whether or not you are dealing with a responsible supplier and investigate alternative sources of supply. However, in many cases a 'stonewall' approach will be received from a supplier who is well aware that he has a unique product which cannot be sourced elsewhere.

In addition to known sources of potential allergens it is necessary to eliminate or control adventitious contamination from other ingredients. A biscuit factory which produces peanut cookies as well as plain will need to take appropriate steps to safeguard the purchasers of the latter variety. The level of control exercised is likely to depend upon a number of factors, not least of which is the size and resources of the operation. A large factory with the production volumes to justify it could address the problem by the use of separate production lines with dedicated raw material handling facilities. A smaller enterprise may rely on the segregation of raw materials with line cleaning taking place after the production of the nut product. It should be emphasised, however, that although the latter solution may be simpler and cheaper, its overall objective must be the same, that is, to eliminate the possibility of cross-contamination from products which do not purport to contain nuts. Such an approach would require a thorough understanding of the line itself together with all of its associated equipment in order to be aware of specific areas where nut traces could be harboured. This is likely to involve a certain amount of line stripdown with careful selection of cleaning methods in order to ensure that all traces are removed and that the cleaning operation itself is not responsible for cross-contamination of an adjacent line. The use of compressed air, for example, should therefore be limited in order to prevent this. Rapid diagnostic techniques for detecting nut presence are available and would prove their worth in demonstrating that the cleaning operation had been successful. After cleaning, the line should be positively released back to production by a suitably senior member of staff. The procedures for carrying out this operation should be documented in order to demonstrate the existence of the system. Records of cleaning should be kept which are signed by the person responsible for the cleaning in order to confirm that it has actually been carried out. The records should be audited and countersigned by quality assurance staff to provide a measure of validation.

Information about potential allergens present in products should be stored in a readily retrievable form in order to facilitate accurate labelling and to deal with customer enquiries which may arise. This can be in the form of either manual records, a product database or one of the electronic specification systems now available.

In some cases, the practicalities of factory layout, the range of products and raw materials handled and other factors may make the elimination of cross-contamination unachievable. In such cases clear labelling of the presence of traces which have the potential to provoke an allergic reaction can provide an alternative approach. With products containing nuts the stakes clearly are higher. In this case, consideration should be given to the manner in which the information is communicated. In some instances, the presence of nuts or nut-derived ingredients is essential to provide the authentic characteristics of the product. In this case, merely labelling their presence may not be enough and it may be necessary to emphasise their presence. This may be done by emboldening the nut ingredients in the list of ingredients. Even this may not provide complete peace of mind, as situations have arisen where a consumer allergic to, for example, almonds but not peanuts has innocently purchased a product clearly labelled as containing peanuts only to suffer an allergic reaction as the product had been produced on the same production line as another product containing almonds. In this case an additional warning in the form 'Warning: this product may contain traces of nuts other than peanuts' will help to overcome the problem. Where it is not possible to produce a product free from nut traces, a suitable warning statement will be needed. Such a warning needs to be bold, concise and compelling in order to have any effect. Care therefore needs to be exercised when selecting the wording, positioning, font and colour for such a statement. On a crowded label this is not always easy, but if it is buried amongst a mass of other text it may fail to deliver its message, with potentially disastrous consequences. Generally, any warning statement should start with the word 'WARNING' in block capitals and be preceded and followed by a clear line of space. Siting of the warning is also crucial: an area of label used for other mandatory text is preferable, as this is likely to be the part of the label where the consumer is looking for information.

Retailers' codes of practice will provide a worthwhile source of advice on the implementation of Good Manufacturing Practice in order to control potential allergens within the manufacturing environment. They may contribute to the 'due diligence' defence, but this will depend upon several factors. First of all, it should be remembered that they are primarily designed to support the retailer's 'due diligence' defence rather than the manufacturer's. Secondly, they should be constructively evaluated by the manufacturer to ensure that they meet his needs. Additional controls may be required to enable them to be fully effective. Lastly, they should actually be implemented and operated; the existence of a code of practice which is not implemented may actually damage the defence.

As indicated, the 'due diligence' defence will apply to offences committed under the Food Safety Act and also under the General Product Safety Regulations. In order to put in place measures to avoid conviction under these Regulations, it is necessary to carry out a risk analysis of the product. This should be a combination of a HACCP (hazard analysis and critical control points) approach and examination of risk using techniques such as brainstorming and lateral thinking in order to identify risks that may not be immediately obvious. By way of an example, consider the situation of a company which produces wine glasses. No doubt the glasses will perform satisfactorily when used for their intended purpose, but what if they are subjected to moderate consumer abuse? Is it reasonable for them to be used for liqueur coffee? This is a foreseeable risk, and should the glasses not be suitable for this purpose and possibly shatter when hot coffee is poured into them, it is suggested that they should carry a prominent warning that they are not suitable for use with hot liquids. In the absence of statutory labelling requirements, thought will need to be given to what information about the presence of potential allergens needs to be provided, how it will be communicated and what training needs to be given to staff. The lateral thinking will come into play when the nature of the service provided is not so straightforward as in a self-service café where labelling or warning notices can be employed. For example, where a catering company services banquets and dinner parties, the guests may not see the menu and the hosts or arrangers of the function may not be aware of any specific allergenic conditions which may affect their guests. In this case some means need to be found of bridging the information gap between the two sides.

It should be noted that the 'due diligence' defence will not provide protection against civil action under Product Liability legislation, although the fact that such precautions have been taken should reduce the opportunity for things to go wrong in the first place.

2.9 The future

Food law is a dynamic entity driven by changing technologies and consumer needs. As such it is inevitable that labelling law will seek to try to meet the needs of those who require information about a product in order to avoid certain medical conditions. This need has been recognised on a global basis: Codex Alimentarius has proposed that all compound ingredients which constitute more than 5% of a finished product (as opposed to the present level of 25%) should be fully declared on food labels, thus revealing the hidden secrets of their potential allergens. It is questionable whether or not this change will actually help the situation for sufferers, as there is at least as great a possibility of the potential allergen being in the undeclared 5% as in the undeclared 25%. The European Commission is considering whether or not this approach should be adopted within Community law. The Commission is also considering an amendment of the Food Labelling Directive to disapply the exemptions to ingredients which contain recognised allergens. The allergens so far proposed for inclusion are as follows:

- Cereals containing gluten and products of these
- Crustaceans and products of these
- Eggs and egg products
- Fish and fish products
- Peanuts and products of these
- Soyabeans and products of these
- Milk and milk products (lactose included)
- Tree nuts and nut products
- Sesame seeds
- Sulphite at concentrations of at least 10 mg/kg.

Other than the requirement on the declaration of starches containing wheat gluten, this initiative is one of the first to specifically recognise and address the question of food allergies.

In order to make food control more proactive, some countries are establishing bodies with specific responsibility for food safety and standards. France has set up the Agence Française de Sécurité Sanitaire des Aliments (the French Agency for Food Nutrition) to assess health and nutrition risks, and the UK is in the process of establishing the Food Standards Agency which will be independent from the Ministry of Agriculture, Fisheries and Food and the Department of Health. Both of these agencies have a wide remit which, although not initially targeting food allergenicity, would not preclude them from doing so in the future.

Outside legal control, UK food retailers have done much to champion the cause of food intolerance. Most major supermarkets require their suppliers to provide detailed information about what is in their own-label products. Such information is available to callers and published in leaflets, and there are proposals to make it available on the Internet.

One of the problems with changes to food labelling legislation is that new requirements inevitably add to the information which the manufacturer must include on the label. In the long term the law of diminishing returns will apply as it becomes more and more difficult for the customer to make any sense of the sheer amount of information on a label. The past few years have seen requirements for information about the presence of genetically modified organisms, the presence of sweeteners, the laxative effects of polyols, and quantitative ingredients declarations. All of this information is competing for space on labels which themselves are a finite size. Add to this the demands of marketeers who have the job of promoting and selling the product and it can be seen that there will soon come a time when there will be no further space to add extra information to a label. This problem has been recognised by EFLA, the European Food Law Association, which represents the views of enforcement officers, retailers, manufacturers and consultants. EFLA have suggested that other means of disseminating information should be explored. Amongst the suggestions are greater use of freephone helplines which can deal with customer enquiries and the use of in-store computers to enable the customer to access information about the products on sale. With the potential for food shopping on the Internet, useful product information could easily become a pre-shopping feature by allowing the customer to access the retailer's food intolerance database at the touch of a button.

2.10 Summary

In this chapter we have seen something of the difficulties which face customers who may have an allergic reaction to a particular food ingredient. We have also seen something of the difficulties faced by food manufacturers in meeting the growing clamour for a greater variety of prepared foods at lower cost whilst trying to safeguard the interests of allergy sufferers. It would be impractical to eliminate from the diet food ingredients which are perfectly harmless to the vast majority of the population. The way forward therefore lies in the provision of good information about exactly what is in each food. Labelling exemptions which provide some flexibility and saving on label space may prove to be a trap for the unwary, as they may mask the presence of ingredients which, even in small quantities, may provoke an allergic reaction. Worse still is the situation where food is sold loose, packed on the premises or sold at a catering outlet. In this case the customer is likely to see little in the way of ingredient information. The key to this is the availability of information, whether by means of product information sheets, staff knowledge or through electronic storage and retrieval systems.

3

Diagnostic tests

B. J. Bateman, The David Hide Asthma and Allergy Research Centre, Isle of Wight

3.1 Introduction

3.1.1 The diagnostic pathway

Diagnostic tests for food allergy, as with all medical tests, cannot be discussed in isolation. They are only one part of the whole diagnostic pathway. When an individual presents a particular problem to a health professional, a diagnostic pathway is embarked upon. This pathway starts with the professional taking the individual's medical history, the story of their particular problem. This is often complemented by an examination. The pathway may or may not conclude with particular tests. All diagnostic tests should be seen within the context of this pathway. Tests only serve to add further pieces of information to that already gleaned from the history and examination. They very rarely alone give a definitive answer.

Using the term 'pathway' makes the process sound linear. It is in fact a cyclical process. Just as the history and examination help the clinician to select relevant tests, information gleaned from diagnostic tests may make the clinician return to the patient, and seek further information from the history or examination.

This chapter will start with an explanation of how to judge any test's 'worth'. It will then describe particular aspects of the history and examination relevant to the diagnosis of food allergies and intolerance. There will then follow an introduction to the wide range of tests used by the clinician in the diagnosis of food allergy, their scientific basis, and evidence of efficacy or utility in the diagnosis of food allergies. Some of these tests are readily available, for example, to general practitioners; others are used only by practising allergists within specialist clinics or even only within research departments. Some of the tests have good evidence to back up their use, whereas others do not. They include tests that depend upon studying the patient's physiological reactions to particular stimuli. These are often referred to as *in vivo* tests and include skin prick tests, patch tests and bronchial provocation tests. Other tests are laboratory-based, *in vitro* tests. These look for specific biochemical markers within the patient's serum, or changes within biopsies of the patient's tissue. There follows a discussion of food challenges and the various outcome measures used to assess whether a food challenge can be defined as positive.

Many 'mainstream' health professionals are involved in the diagnosis of food allergies and intolerances. These include medical doctors from many disciplines. All branches of nursing will encounter at some stage a patient with a food allergy. The chief allied profession involved is the dietician, whose role is essential in both the diagnosis and ongoing management of people with food allergies.

The complementary or alternative healthcare system has another group of workers involved in the diagnosis and treatment of those who may have food allergies. Many therapists are members of bodies that ensure careful control of training and practice. Some of their diagnostic tests, however, lack evidence to support their use. There are also individuals with little or no training, however, and with less transparent credentials selling schemes purporting to diagnose food allergy, either by mail or through high street healthfood shops.

3.1.2 The accuracy of diagnostic tests

Every disease has a rate of prevalence and incidence within both the general population and specific populations. The term 'prevalence' is a statistic based upon a particular point in time. It refers to the number of cases of a particular disease divided by the total number of people within the population and is usually represented as a percentage. 'Lifetime prevalence' is the number of people within a population who may have a particular disease at some time in their life, expressed as a percentage of the total population. The term 'incidence' refers to the number of new cases of a disease occurring over a specified period of time. The two terms are useful for different kinds of disease. The prevalence of a disease is often useful for more chronic diseases – those diseases which people rarely recover from, but also rarely cause death. A useful example is an estimate of the lifetime prevalence of peanut allergy within a given population. Diseases with high recovery rates or with high mortality rates are more usefully explored using the concept of 'incidence'. These include childhood infectious diseases. They also include food allergies that commonly occur in infancy and are known to have a high rate of resolution, such as milk or egg allergy.[1]

Obtaining accurate estimates of incidence and prevalence of diseases is not always simple. There are problems not only in defining true cases of the disease but also in making good estimates of the total population. From the available figures of incidence and prevalence, either formally or informally, the health professional develops an idea of the 'risk' that a particular patient has of a particular disease. This may simply be the risk derived from the general population's incidence or prevalence of the disease. If a chronic but not usually life-threatening disease has a prevalence of 1%, a particular patient has a risk of one in a 100 of having the disease. Further refining the prevalence from the patient's particular sub-population will modify each assessment of 'risk'. If the patient is female, and the disease has a 2% prevalence within the female population, her risk will increase to two in 100.

Each piece of new information, whether gleaned from the history, the examination, or subsequent diagnostic tests, progressively modifies the individual's risk of a given condition. Each individual can be thought of as having a prior or pre-test probability of the illness. Subsequent to a positive or negative result from a particular test, a posterior or post-test probability can be calculated. A positive diagnostic test will have different implications for an individual whose risk of food allergy has already been estimated as high from their history, compared to those for an individual drawn from the general population, with an unspecified but certainly lower risk. A patient who is seen in a non-specialist clinic is likely to have a lower risk than a patient seen by a specialist, as a filtering process will already have taken place. The impact that the test has on each individual's risk can be expressed statistically as a likelihood ratio.

The likelihood ratio can be calculated from the test's sensitivity and specificity.[2] Consider a given disease in a population, such as peanut allergy. There will be people who definitely have this disease, who have immunologically mediated reactions to peanut proteins. One needs to identify a 'gold standard' against which to measure the performance of other tests. This 'gold standard' is the best method available for estimating the prevalence of people who really have immunologically mediated reactions to peanut proteins. Let us consider the double-blind placebo-controlled challenge as our gold standard. At present within the field of food allergy it is probably the nearest we have to a 'gold standard'. If a patient has a clinically documented reaction by a blinded observer to hidden peanut protein, they are regarded as having the disease known as peanut allergy. We can then assess various tests against this 'gold standard' as to their efficacy at identifying true positives and true negatives.

How good is each test at correctly identifying those patients who truly have the disease, and at correctly identifying those who do not?

A highly sensitive test is one that is very good at identifying all cases of the disease while also including many of the normal population. A negative result is particularly useful, as the person showing that result is very unlikely to have the disease. Sensitivity can be expressed mathematically as a ratio, calculated thus:

$$\text{Sensitivity} = \frac{\text{true positives}}{\text{true positives} + \text{false negatives}}$$

Specificity, however, is a measure of how important a positive result is. A positive result from a highly specific test is very likely to indicate that the individual showing that result has the disease, whereas a negative result does not so reliably rule out the condition. This can be expressed mathematically again as a ratio, calculated thus:

$$\text{Specificity} = \frac{\text{true negatives}}{\text{true negatives} + \text{false positives}}$$

A test often forfeits specificity at the expense of sensitivity. The most desirable test is both highly specific and highly sensitive, but in different situations sensitivity might be more important than specificity, while others one may be more interested in the specificity of the test.

3.2 The clinical history and examination

The clinical history and examination is the basis of the medical assessment. Many of the professions allied to medicine use a similar starting point. It is the first stage of risk assessment. The clinician is using the information she or he gathers to modify the patient's risk from that of the general population to a more specific one. The history taking will be considered under the headings commonly employed by physicians:

- Presenting complaint
- Past medical and drug history
- Occupation and smoking history
- Family history
- Examination.

3.2.1 Presenting complaint

The clinician takes a detailed account of the patient's presenting complaint, initially using open questions, and then using closed questions to obtain important details. The information gathered includes the timing and frequency of symptoms, any precipitants that the patient may suspect and any adjuvants such as alcohol or exercise. The clinician is looking for common or recognised patterns of symptoms. The focus of the questioning will differ depending on the age of the patient. The clinician concentrates upon symptoms that have been confirmed to occur as a result of food allergies. Symptoms are more easily dealt with if considered on a system-by-system approach. Common patterns of symptoms arising from food allergy occur within the gastrointestinal system, the skin, the respiratory system and the cardiovascular system.[3] Patients commonly have symptoms in different systems. The common symptoms are listed in Table 3.1.

Common symptoms within the gastrointestinal system include nausea and vomiting, diarrhoea, abdominal cramps, bloating and gas. The symptom of infant 'colic' is also often included within this system. Symptoms associated with the skin include pruritus (itching), atopic dermatitis/eczema, urticaria or hives, erythema (redness) and angioedema (swelling of the face, lips and

Table 3.1 Common symptoms associated with food allergy

System	Common symptoms
gastrointestinal system	nausea and vomiting diarrhoea abdominal cramps bloating and gas infant 'colic' (often included within this system)
skin	pruritus (itching) atopic dermatitis/eczema urticaria/hives (rash like 'nettle' rash) erythema (redness) angioedema (swelling – face and mouth)
respiratory system	sneezing eye inflammation and tears serous rhinorrhoea (runny nose) laryngeal oedema (swelling of voice box) cough bronchorrhoea (mucus production) wheeze breathlessness
cardiovascular system	tachycardia (raised pulse rate) arrhythmias (abnormal heart rhythms) hypotension (fall in blood pressure)

mouth). It is important for the clinician to note how long the symptoms have persisted. If the patient gives a history of *acute* urticaria (defined as occurring for less than 6 weeks), the symptoms are more likely to be due to foods than if they have been present for longer, that is, if the patient has *chronic* urticaria.[4] Interestingly, although skin symptoms are a common manifestation of food allergies, they do not seem to be a prerequisite for the diagnosis of serious adverse food reactions.[5] The respiratory system is rarely affected without signs or symptoms within one of the other systems. Symptoms include sneezing, evidence of ocular inflammation, including injection (visibility of the surface blood vessels) and tears, serous rhinorrhoea, laryngeal oedema, cough, bronchorrhoea and evidence of bronchoconstriction, such as wheeze and breathlessness. Symptoms within the cardiovascular system are rare without signs of adverse food reactions in other systems. Cardiovascular symptoms include tachycardia, arrhythmias and hypotension.

The history should also include questions regarding symptoms of other atopic diseases. This 'family' of diseases includes asthma, eczema and atopic dermatitis, and perennial and seasonal rhinitis (hayfever). Patients who have one atopic disease are known to be more at risk of IgE-mediated food allergy.[3]

People do report less common symptoms associated with food. These have been discussed in Chapter 1. The potential mechanisms for such symptoms are controversial, and are unlikely to be typical IgE-mediated/Type I hypersensitivity reactions. Type II reactions, antibody-dependent cytotoxicity, or Type III antigen–antibody complexes may be implicated.[3] Some are recognised, if rare, often without full understanding of the immunological mechanism. Some symptoms lack convincing proof that they are attributable to immunological mechanisms.

Food is causally linked with symptoms of rare diseases, some with serious complications. The ingestion of cow's milk has been linked, if rarely, to the bruising and petechiae (pin-prick rash under the skin) associated with thrombocytopenia (low platelet count with resultant problems with clotting).[5]

There are several problems that are specifically associated with the gastrointestinal system of infants. Some are more serious than others. Infants may become profoundly unwell with food-induced enterocolitis and enteropathy (inflammation and malfunction of the digestive tract).[5] The rectal blood loss associated with food-induced (usually cow's milk or soya) proctitis (inflammation of the rectum) causes great concern to parents but fortunately is rarely associated with any other symptoms and is almost never sufficient to cause anaemia.[5]

Coeliac disease has a clearly described pathophysiology, with a wide range of symptoms and its own set of diagnostic tests. It results from a permanent sensitivity to gluten, a protein present in wheat and oats. This sensitivity is immunologically mediated (a Type IV reaction rather than the Type I, IgE-mediated reactions, which are much more commonly associated with food allergy). The sensitivity results in an enteropathy, the architecture of the wall of the small intestine is progressively distorted and inflamed and the bowel stops digesting and absorbing other foods efficiently. This is the cause of many of the symptoms, which are extremely varied. They include gastrointestinal symptoms such as diarrhoea and vomiting, or nutritional deficiencies leading to weight loss or failure to thrive in children. The behaviour of both children and adults can be affected and patients have been known to present with depression. Individuals with a particular genetic makeup, with the human leucocyte antigen (HLA) type A1, B8 DR3/7 are more at risk of each of coeliac disease and other associated autoimmune diseases (diseases where the body produces antibodies against some of its own parts).[6]

Claims that foods may cause or exacerbate other symptoms than these have been made by patients and clinicians alike. There is little convincing evidence to support a direct causal link, mediated by a recognised immune response in the area of behavioural changes, migraines and epilepsy. There may be a pharmacological mechanism with vasoactive compounds present in a variety of foods, such as coffee, cheese, chocolate and red wine.[7] Similarly there does not appear to be good evidence to support any causal links with food and inflammatory bowel disease (Crohns's disease and ulcerative colitis).[8] There is some case-study evidence associating food reactions with the exacerbation of arthritis.[9]

3.2.2 Past medical and drug history

The patient's full past medical history, including drug history, should be documented, including other allergic and all non-allergic illnesses.

The possibility of a psychiatric history should be considered. Some clusters of symptoms at presentation are more likely to be linked with psychiatric diagnoses. People presenting with multiple symptoms, and concerns over many foods and other environmental problems, have been shown to be more at risk of symptoms of depression or anxiety.[10] Parents may make claims of multiple food allergies in their children. Such claims have been known to be sufficiently extreme to be diagnosed as Munchausen's by Proxy.[11]

The assessment of the patient should include a drug history. This will aid the identification of drugs that may be the cause of the patient's symptoms, as in the association between urticaria and aspirin or between asthma and other non-steroidal anti-inflammatory drugs.

Beta blockers are said to make the possibility of a severe anaphylactic reaction more likely, and may modify the advice and emergency treatment the clinician gives to the patient.

A drug history will also enable the clinician to optimise the patient's therapy, and ensure they have the necessary emergency drugs in their possession and are confident in the timing and method of their use.

3.2.3 Occupation and smoking

Respiratory diseases have known associations with those working in the food and food-related industries. These include occupational asthma, occupational rhinitis and hypersensitivity pneumonitis. Skin diseases such as contact dermatitis and contact urticaria are also associated with work in these industries.

These diseases are not all Type I, IgE-mediated reactions. Some cases of occupational asthma and some of contact dermatitis occur as a result of irritation.[12] Hypersensitivity pneumonitis occurs as a result of a Type III or possibly a Type IV hypersensitivity reaction.[13] As with non-industrial food allergy or intolerance, the pathophysiological mechanism affects the choice of diagnostic tests.

All histories should include questions about the individual's employment and hobbies. If the pattern of disease makes the clinician suspect a temporal relationship with work, they should seek details about recent changes of workplace and recent changes in tasks and the introduction of new materials and processes.

The temporal relationship is not always obvious. Some individuals suffer from a late bronchial reaction to the agent, with the decline in lung function seen only at 2–10 hours post exposure and not returning to normal until sometimes 36 hours later. A latent period between exposure to the allergen and subsequent development of the symptoms is usually reported. As with all allergic diseases, the individual requires a period to acquire sensitisation; this can be anything up to 20 years after the initial exposure. Workers in particular industries are known to be at higher risk of occupational asthma. More than 50 agents have been implicated. Such diseases among workers in food industries are sometimes due to the food proteins themselves, but in other cases the specific IgE is directed against micro-organisms or insects involved in food production or storage. Workers at risk include those in contact with wheat and rye flour, green coffee beans and snow crab. Workers who handle mushrooms and celery can be sensitised either to the food proteins or to micro-organisms associated with these vegetables.[14]

The diagnosis of work-related illnesses has important implications for the individual's livelihood, often necessitating permanent removal from exposure to the agent. It may also have important legal consequences for the employer.

It appears that smoking may well increase the risk and speed of sensitisation to inhalant allergens, making smokers more at risk of IgE-mediated illness when exposed to sensitising agents, particularly in the workplace.[15]

3.2.4 Family history

The propensity to develop atopic illnesses appears to be partly genetically determined. Although quantification of the genetic risk is a complex issue, there seems to be approximately two to three times the risk of having an unspecified atopic illness if one has a relative with an atopic illness.[16]

3.2.5 Examination

A general examination of all the major systems of a patient normally supplements the clinician's history taking. In the case of paediatric examination, in particular, it should include the measurement of parameters of growth – height, weight and head circumference – which should be recorded on an appropriate centile chart. Chronic illnesses such as coeliac disease and poorly controlled asthma may result in a thin, short child, as do the use of long-term high-dose steroids.

The examination of the patient suspected of having an allergy to a food then focuses upon the presence or absence of signs associated with other atopic diseases. This not only establishes whether the individual has an atopic disposition, but also may identify signs resulting from, or exacerbated by, their food allergy.

The signs of eczema and atopic dermatitis have proved difficult to define.[17] The acute signs include erythematous (red) and vesicular (blistered) skin. More chronically one finds lichenified (thickened), oedematous (swollen) and cracked areas of skin. The picture and distribution is slightly different depending upon the age of the patient.

The physical examination of a patient who lists urticaria and/or angioedema as their symptom is often unremarkable. There should be a particular emphasis on the search for the signs of other systemic illnesses known to be associated with urticaria. Any urticarial lesion should be noted, such as an itchy well-demarcated raised area, often with surrounding erythema.

Patients with perennial or seasonal rhinoconjunctivitis may have injected conjunctivae (visible small blood vessels), erythematous conjunctivae (reddened whites of eyes), puffy eyelids, and erythematous, oedematous nasal mucosa (the lining of the nose appearing swollen and red). Studies investigating any link between food allergy and otitis media with effusion (sometimes termed glue ear – long-standing fluid in the middle ear resulting in, albeit temporary, conductive hearing loss) have been poorly conducted.[18] To date, there is no good evidence linking this condition with food allergy.

Asthma is more usually diagnosed by history and if necessary confirmed by lung function tests. Examination may rarely reveal the acute signs of respiratory distress and wheeze associated with asthma, or more chronic changes, such as Harrison's sulci (a change in shape of the lower rib cage).

3.2.6 Most common foods

When taking the initial history of a presenting complaint, the clinician will try to elicit whether the patient suspects one or more foodstuffs. As discussed in Chapter 1, there is a limited list of foods that cause the majority of reactions. Patients, however, may not be aware of which specific food caused their reaction; there are many cases of reactions resulting from the consumption of food contaminated with only traces of the harmful ingredient.[19]

Strawberries, other berries, tomato and citrus fruits are commonly reported as producing a flush or rash, particularly on the faces of young children. This phenomenon does not appear to be an IgE-mediated reaction but may have a pharmacological basis.

The oral allergy syndrome discussed in Chapter 1 solely involves the oropharynx (mouth, tongue and throat). Patients describe the rapid onset of itching of the mouth and angioedema (swelling of the lips, tongue, palate and throat). This is generally followed by a rapid resolution of symptoms. They are most commonly associated with the ingestion of various fresh fruit and vegetables. Patients with allergic rhinoconjunctivitis ('hayfever') associated with airborne allergens are most commonly afflicted with this problem. Care must be exercised when taking the history that these symptoms were not in fact the herald of more generalised systemic symptoms.

3.3 Diagnostic tests

Diagnostic tests can be divided into *in vivo* tests and *in vitro* tests. *In vivo* tests involve the patient themselves. The gold standard of all these is the double-blind placebo-controlled food challenge (DBPCFC). Outcome measures from DBPCFC may include patient reports of symptoms and clinician observations, or more sophisticated tests such as pulmonary function tests. Other *in vivo* tests include assessment of the presence of specific IgE, using skin tests. Assessment of late-phase reactions may rarely be aided by the use of patch tests.

In vitro or laboratory-based tests mainly fall into two groups, those which can be carried out independently of any recent reaction and those dependent upon a recent reaction. The first group are those which aim to identify the presence within serum of specific IgE. The second group of tests are at present largely confined to the research arena.

3.4 Food challenges

People seem to be very ready to attribute many symptoms to either a specific food or a range of foods. Studies suggest that at most 50% of patients suspected of having a food hypersensitivity will have a positive double-blind placebo-controlled food challenge.[20] This statistic is from selected populations derived from allergy clinics. These people are almost certainly at higher risk of definite food allergy than an individual selected from the general population in a more general clinic, because of prior 'selection' based upon history and examination. The 'placebo effect' of many foods is strong and has been demonstrated in many studies comparing the results from open (unblinded) tests with those from blinded ones.[20] The placebo effect is applicable to both patient and investigator alike, underlining the importance of blinding both investigator and patient.

The DBPCFC is largely restricted to practising allergists, and not widely available in the United Kingdom. Paediatricians in the UK more commonly use open challenges. This is appropriate in infants where there is less of a problem of psychological overlay. There is still the issue of the psychology of the parents and practitioner. A negative open challenge is very useful in ruling out the role of food in causing or exacerbating the symptoms.

There is also a therapeutic as well as a diagnostic role for food challenges. This is best illustrated in children. Infants with food allergies, in particular allergies to egg and milk, have a high likelihood of becoming tolerant to the food, 'growing out' of their allergy. Interval re-challenges are essential, to 'test' their continuing sensitivity to the particular food.

The designing of a food challenge involves several steps that are detailed below.[21]

3.4.1 Careful history

History gathering, as discussed earlier, should concentrate on the most likely foods and symptoms. Clearly some patients will not be suitable for a food challenge. Some patients will not be prepared to have what may be strongly held beliefs investigated, and the history should include some assessment of the suitability of the patient for the procedure. Other pieces of additional information essential for the design of the challenge include the timing between the ingestion of the food and the onset of symptoms, the amount of food necessary to produce symptoms, and finally any adjuvant factors such as exercise that are necessary for the onset of symptoms. Nearly all the foodstuffs positively incriminated cause symptoms within hours of ingestion (except protein-sensitive enteropathies). Some patients may give very specific symptoms that have measurable parameters, useful as outcome measures of the DBPCFC. Respiratory symptoms can be monitored using pulmonary function tests and bronchial provocation challenges. Peak expiratory flow rate (PEFR) and the forced expiratory volume in one second (FEV1) are the most reproducible measures affected during bronchoconstriction (narrowing of the airways). Specific bronchial provocation challenges can be useful in the diagnosis of food allergies. These have been employed particularly in the confirmation of occupational or industrial asthma. Specific bronchial challenges, using aerosol preparations of the implicated allergens, can be used to demonstrate resultant bronchoconstriction.[22]

3.4.2 Assessments of specific IgE – via skin prick tests or *in vitro* methods

The use of skin prick tests and *in vitro* identification of specific IgE is discussed later in this chapter. Such tests are usually insufficiently sensitive or specific to be used in isolation for the diagnosis of food allergies. It is suggested that the only reason for not proceeding to DBPCFC is if there is strong suspicion that a likely food substance caused an anaphylactic reaction, and positive evidence of specific IgE. Open challenge and if necessary DBPCFC should follow negative skin prick tests. Positive tests in the presence of non-life-threatening symptoms should be followed with a DBPCFC.

3.4.3 Elimination diet

The patient should be established in a stable symptom-free, or minimal symptom, period. This is achieved through either an extensive elimination diet, limited to only a few foods, or a simpler one eliminating only the suspected foodstuff. The use of a dietician is invaluable here, to ensure the nutritional adequacy of the elimination diet, particularly when investigating children.

3.4.4 Open challenge/reintroduction of normal diet

There may be no need to progress to DBPCFC. The reintroduction of a normal diet should be considered in the patient whose symptoms appear to have been only minimally altered by an adequate elimination diet. If this is then followed by the relapse of symptoms, the patient requires reassessment, reapplication of the elimination diet and probable progression to the DBPCFC. Single-blind challenges, where the investigator is aware whether the consumed food is placebo or active, if negative, are useful. They sometimes enable the clinician to avoid having to proceed to DBPCFC that is more resource-intensive to organise. The positive single-blind challenge should ideally be confirmed by a double-blind challenge.

3.4.5 Choice and preparation of foodstuffs, challenge vehicle and placebo

Neither the patient nor observers should be able to distinguish between the active and placebo challenge: 'blinding' needs to be adequate. To this end, active food needs to be hidden either within opaque capsules or within other strongly flavoured food vehicles to which the patient is known not to be sensitive.

3.4.6 Challenge (double-blinded)

Challenges progress through increasing doses of the active foodstuff. In practice, investigators start at half the minimum amount that the patient has reported as causing symptoms and then double the amount at each step. The time period between each challenge should be greater than the time interval within which symptoms occur. The patient should be carefully observed throughout the challenge procedure, or if the challenge is occurring at home they should keep a symptom diary. Some advise a separate placebo challenge period with similarly increasing amounts of placebo matching the active challenge. The order of the placebo and active challenge should be randomly decided. The investigator may need to carry out several challenges, in particular if the patient suffers mainly from subjective symptoms, with no more objective signs.

3.4.7 Open consumption of the food in usual circumstances

To complete a negative challenge it should be followed by open consumption of the foodstuff, prepared in a normal way, consumed in normal amounts with necessary adjuvants.

3.4.8 Safety of DBPCFC

The risk of generalised anaphylactic reactions should be considered when undertaking a challenge with a foodstuff. British practitioners tend to work from within a hospital setting, while Continental and American allergists often work from 'stand alone' offices or clinics. Whatever the context, there are good guidelines on the personnel and equipment necessary for those undertaking immunotherapy using allergy extracts.[23] These standards are also applicable to food challenges. General principles of resuscitation apply, with the ready availability and if necessary administration of subcutaneous epinephrine/adrenaline as the first-line drug in the case of a systemic reaction. Antihistamines are only useful adjuvants, their mode of action being too slow in this scenario. Two North American allergists, S. A. Bock and H. A. Sampson, who have performed many food challenges and written extensively on the subject, state that they have never had to use intravenous resuscitation, and have never had an episode of cardiopulmonary arrest.[24] Prudence, however, dictates that a patient with a history of anaphylaxis should be challenged only in the hospital setting with very small incremental doses of food.

3.5 Skin testing

Skin testing is an *in vivo* method of identifying the presence in an individual of specific IgE to a given allergen.[25] It is useful for both clinical practice and research purposes. Clinically it is used to supplement the patient's history and examination. Reactions to specific allergens will guide the clinician as to the specific substance that has caused or is causing the patient's problem. If the patient had a life-threatening reaction and the results of the skin prick test are concordant with the patient's clinical history, the clinician and patient may opt to take no further diagnostic action, and simply avoid the particular foodstuff. More commonly, however, the clinical history and skin test results are not sufficient. There may be positive skin tests without a confirmatory history, or a suspicious history with negative skin tests. This calls for the use of the double-blind placebo-controlled challenge to confirm or negate any suspicions that may have arisen from the history and examination or skin tests, in particular to prevent the potentially nutritionally and certainly socially harmful effects of an unnecessarily restricted diet.

Skin tests are simple, readily available, rapidly performed and inexpensive. They have the disadvantage of producing a significant number of false positives and false negatives, especially when improperly performed. They rely upon the presence within the skin of all the necessary cells and mediators for the occurrence of a Type I hypersensitivity reaction. A small amount of allergen is introduced percutaneously. An immediate weal-and-flare reaction (see below) is provoked in the patient's skin. This is dependent upon proinflammatory and neurogenic mediators. Histamine and tryptase are released by mast cell degranulation after the introduction of allergen, recognised by specific IgE. Histamine is the major mediator of the weal-and-flare reaction. The immediate reaction occurs about 15–20 minutes after the test, and is used as evidence of a positive reaction. Visual evidence of this reaction is oedema (the swollen *weal*) and erythema (the surrounding red *flare)*. The patient will often also complain that the area itches.[25]

There are two main methods of performing skin tests: the prick-puncture and the intradermal method. The main issues surrounding their use are their diagnostic utility, and their respective levels of comfort and safety.

3.5.1 Prick-puncture tests

Prick-puncture tests are less sensitive and less reproducible, but more specific than intradermal tests.[26] The specificity of prick tests and their superior safety profile is the reason why they are recommended by the European Academy of Allergology and Clinical Immunology and the US Joint Council of Allergy, Asthma and Immunology.[25]

Each antigen is placed upon the skin and introduced into the epidermis. The test is most reliably carried out on the volar aspect of the forearm, though occasionally the patient's back is used if a larger area is needed or the skin of the arm is affected by eczema. Small drops of each allergen extract are placed at least 2 cm apart on the skin. A hypodermic needle is placed at an angle through the drop of liquid and the needle tip is gently inserted to lift the top layer of epidermis, without causing bleeding. A separate needle is used for each allergen extract to reduce the risk of cross-contamination. Positive and negative controls are also used. The negative control identifies the patient with significant dermographism (non-specific skin sensitivity), reducing the chance of false positive reactions. This is commonly allergen diluent, or normal saline. The positive control commonly used is histamine phosphate (1 mg/ml of histamine base).

Other prick-puncture test methods have been developed. These involve the introduction of specific instruments perpendicular to the skin, such as the Morrow Brown standardised needle. This is an attempt to improve repeatability of the test. Opinion is divided as to which method is preferred, although some guidelines do recommend the perpendicular method. [25]

Prick-puncture is a safe test; there have been very few reports of systemic reactions, and no fatalities.[25]

3.5.2 Intradermal test

The intradermal test involves the intracutaneous injection of 0.01–0.05 ml of allergen extract through a 26/27-gauge needle. Similar positive and negative controls are also used. This method carries more risk of systemic reactions than the prick-puncture method. It frequently produces local reactions, making it a less 'comfortable' test. The local reactions also have implications for the accuracy of the test, with an ensuing high rate of false positives. There are reports of systemic reactions, including fatalities. Its safety can be improved by preceding the test with prick-puncture tests, using serial tenfold dilutions of the usual test concentration, in particular if subjects have a history of anaphylactic reactions. The presence of a physician is recommended, with resuscitation equipment including pre-loaded adrenaline.[25]

3.5.3 Interpretation of skin tests

The size of the weal-and-flare should be read at the peak of its reaction, after approximately 10–20 minutes. A copy of the weal-and-flare reaction should be transferred using pen and clear tape, to ensure a permanent record is kept. The mean of the longest and midpoint orthogonal diameter of the weal has been shown to correlate well with more precise planimetry methods, despite the weal often having a rather irregular shape.

The 'cutoff' at which one declares a test positive will influence the test's sensitivity and specificity. A 3 mm mean weal diameter is the common definition of a clinically significant positive reaction, corresponding to a 10 mm mean flare diameter.

Researchers or clinicians may wish to quantify the reaction to assess how much specific IgE the patient has. There is no good correlation between the size of the weal and the amount of specific IgE. In Scandinavia workers compare the size of the weal-and-flare reaction with the size of the reaction produced by the positive histamine control, but this too is inaccurate.

The operator plays a significant role in the reproducibility of skin prick tests, and it has been suggested that duplicate skin tests should be performed. A rate of 5% single-negative tests in clearly sensitised subjects is to be expected, even in the most experienced operator's hands.

The type and quality of the allergen clearly affects the diagnostic efficacy of all assessments of specific IgE, including skin tests. Attempts to typify the specific proteins from each foodstuff that are most commonly responsible for clinical symptoms are ongoing. As these are described they will enable skin prick tests with these recombinant allergens, possibly improving standardisation. Fruit and vegetables seem particularly difficult substances from which to produce reliable allergen extracts. Some workers advocate the 'prick–prick' method, first pricking the relevant fruit or vegetable, and then the patient's skin. The concerns about this method include lack of standardisation of the allergen.

Other factors which may influence the size of the reaction include age (skin test weals increase from infancy to adulthood and then often decline after the age of 50), race (dark skin pigmentation elicits a greater weal response from histamine), season, pathological conditions, and drugs.

Those without symptoms but with positive skin prick tests may lie in one of two groups. They may indeed be false positives, and the positive reaction may be due to irritants or other mast cell secretagogues and not an indication of specific IgE. The other group includes the asymptomatic but skin prick test positive people who are at greater risk of developing allergic symptoms, but not necessarily food allergies, later in life. This is termed 'latent allergy'.[27]

The negative skin prick tests of those with symptoms may be explained by poor technique, drugs or disease attenuating the skin's reaction, poor quality extracts and decreased reactivity of the skin of infants and the elderly.

The interpretation of positive skin prick with food extracts is even more difficult than with aeroallergens. Only a fraction of people even with positive reactions to the more specific prick-puncture tests to foods will react during a challenge.[20]

3.6 Patch testing

Patch testing is a diagnostic tool commonly employed in the diagnosis of contact dermatitis. This may be irritant or allergic in origin. It may be difficult and even artificial to distinguish between these two. Irritants make up about 80% of the problem, and an allergic cause can be attributed to about 20% of patients with contact dermatitis. It is rarely used in the investigation of systemically induced food allergies.[12] Patch testing infants suffering from eczema with cow's milk may hold some diagnostic promise.[1]

Patch testing is commonly used to identify substances to which the patient may have become sensitised and that are either causing their dermatitis or aggravating their eczema/atopic dermatitis. These substances are derived from many sources – environmental, domestic or cosmetic, or the topical applications actually being used to treat the patient's skin. Food substances can cause contact dermatitis.

A standardised battery of substances is placed upon the patient's back within small aluminium chambers, under occlusive dressing. There exist a European and a North American Standard Series. Occasionally more specific series are chosen. They are left *in situ* for 48 hours. The dressing is then removed and the back is wiped free of any residual chemicals. Each specific area of the patient's skin is examined for any signs of dermatitis – erythema, oedema and blistering – and a score of 1+ to 3+ is assigned depending on the severity of the reaction. The score is defined as 1+ if there is erythema and oedema covering at least half the test area, 2+ if there are also papules and 3+ if there are vesicles or bullae present. Each area is then read again between 96 hours and 7 days later, for persistence of the reactions, or for emergence of new reactions. This later reading increases the sensitivity of the test.

3.6.1 Application to the diagnosis of food allergy

Foodstuffs are an uncommon cause of immunologically mediated contact dermatitis, although regular contact with vegetables and meat can certainly irritate intact or eczematous skin. Plant-derived saps such as from poison ivy and poison oak can cause a characteristic contact dermatitis. There exists cross-reactivity with these saps and mango skin and the oil from cashew nut shells.

Some clinicians claim an exacerbation of eczema/atopic dermatitis in patients who are patch test positive to nickel, cobalt and balsam of Peru, following oral ingestion of foods containing these substances. They similarly claim an improvement in these patients' conditions when they manipulate their diet to reduce the amount of these substances. There is some double-blind placebo-controlled evidence to support this, although there are some problems with the study design, in particular with the amount, source and form in which the salts are ingested. An exacerbation of eczema following oral ingestion of foods to which patients are patch test positive is not a commonly accepted view.[28]

3.6.2 Open patch test and the diagnosis of contact urticaria

Some food substances can induce an immediate urticarial-type reaction at the point of contact. No standardised test exists for investigating such contact urticaria, but one can demonstrate such a reaction by an open test. The substance is placed on the skin of the flexor surface of the forearm for 30–45 minutes in an attempt to replicate the urticaria. It may be necessary to use non-intact, eczematous skin. This contact urticaria may be secondary to an allergic or non-allergic reaction. In the non-allergic type no previous sensitisation has taken place; the individual does not have specific IgE to the substance. The urticaria occurs because of non-immunological release of vasoactive substances in the skin. Substances that may affect the skin by this mechanism include acetic acid, benzoic acid, cinnamic acid, sorbic acid and balsam of Peru. Contact urticaria can also be mediated by allergic mechanisms, chiefly specific IgE mediated. Foods capable of causing a reaction in such sensitised people include milk, eggs, fish, nuts, fruits and vegetables.

3.7 Laboratory tests

Laboratory tests fall into two groups. The first are those that identify substances present independent of a recent or concurrent IgE-mediated reaction. The most widely used tests are based upon the serum assays of specific IgE. More direct evidence of the potential for a reaction can be obtained by challenging *in vitro* suspensions of extracted leucocytes (white blood cells) with allergen. Other, non-IgE antibodies can be assayed using similar techniques to those used for the detection of IgE molecules, but their diagnostic relevance is less clear. There are tests that identify cells rather than antibodies. Reports include the detection of specific peripheral mononuclear cells (white blood cells which circulate in the body). This process remains confined to the research arena.

The second group of tests includes those that are dependent upon a recent or concurrent reaction. At present these too are confined to the research arena. They may prove useful in the future in the retrospective diagnosis of reactions to food allergens. On the whole they do not give information about which specific allergen produced the reaction but only supply supportive evidence of a recent immunologically mediated reaction. They include measurements of various cell mediators.

Double-blind placebo-controlled food challenges remain the 'gold standard' in the diagnosis of food allergies, but *in vitro* tests can supplement this.

Nearly all current diagnostic knowledge lies within the area of Type I, IgE-mediated mechanisms of food allergy. This is the only conclusively proven aetiology of food allergy.

3.7.1 Technical validity and clinical validity

The distinction needs to be made between the technical validity and the clinical validity of a test. One may demonstrate the analytical specificity of a test, detection of IgE that will react with particular antigens reliably and reproducibly *in vitro*. This does not necessarily mean that the person whose serum was the source of the specific IgE will have a clinical problem with that food, the clinical specificity of the test. For this the clinician needs information about the test's diagnostic performance when compared against the DBPCFC. This information is often not available. The manufacturer should demonstrate the technical validity of the test by ensuring its *analytical sensitivity*, *analytical specificity*, *accuracy* and *precision*. The performance of the test when measured against the clinical problems of the patient from whom the serum was extracted has been discussed earlier.

3.7.2 Specific IgE

Tests that detect antigens and antibodies use similar immunological principles, relying on the antibody–antigen reaction. They are referred to as immunoassays or immunochemical techniques. All specific IgE molecules belong to one class or isotype of antibody. The other isotypes are IgG, IgA and IgM antibodies. Even though they have variable antigen-binding regions, giving them their specificity, they have an effector area that is constant. Antibodies can be raised, using monoclonal techniques against this constant area, producing anti-IgE antibodies.

The allergen is presented to the serum in a variety of forms. The antibodies within the serum then either react or do not react with the presented allergen. The reacting antibodies are then 'labelled' by the binding of a detection, anti-IgE, antibody to the serum IgE.

Both the appropriateness and standardisation of the allergen are important in this diagnostic test. Clinical validation depends upon sera from individuals with DBPCFC-proven food allergies. This can be problematic when developing assays for the investigation of rare food allergies.

Different methods use different allergen presentation systems (fluid phases, microparticles and various solid phase presentation vehicles such as polystyrene or paper), and different anti-IgE detection 'labelling' systems (radioactivity, enzyme reactions, and enzyme systems linked with fluorescence and chemiluminescence).[29] The radioallergosorbent test (RAST) and the CAP radioimmunoassay (RIA) systems exploit radioactivity as the labelling mechanism, with various antigen presentation systems within different kits. The CAP FEIA system uses the principle of enzymes linked with fluorescence as its labelling system, having presented the antigen on a cellulose 'sponge'.

An international scale for the *in vitro* quantification of IgE has proved difficult to produce. This is because different sera from different allergic patients demonstrate different quantitative responses, in particular as different patients react to different epitopes on each allergen.

Two systems are used in practice for the quantification of specific IgE, though both are prone to error. They do show that a reasonable degree of correlation is obtained between different detection systems. Reports vary concerning the concordance of results from different laboratories and different systems. They range from up to 90% of laboratories with concordant results for common inhalant and food allergens to reports from other authorities of much lower concordance when comparing the various detection systems. Bindslev-Jensen and Poulsen quote figures for different methods of detecting specific IgE to different food allergens. They list high levels of specificity, between 80% and 90% and even approaching 100%, for cod. When they examined specific IgE to cereals, sensitivity was not quoted and specificity was very low, the lowest being 16% for one of the systems for rye. The sensitivity with different systems for hazelnut was only about 50%. As researchers have not been able to discover a 'level of discrimination' above which patients will have clinical symptoms and below which such symptoms are unlikely to occur, manufacturers have opted for high sensitivity at the expense of specificity. It is clear from these comparisons that a particular system cannot be recommended for all food allergens, as the results of each test need to be taken in the context of both the particular allergen and the system used to identify the specific IgE.[30]

It is clear that some allergens share common epitopes (areas of the protein recognised by each specific IgE molecule). This has implications for both clinical cross-reactivity and cross-reactivity *in vitro*. Specific IgE may bind with a related allergen with a common epitope. If a patient has clinical problems with a particular allergen, what is the likelihood that they will have problems with other allergens sharing these common epitopes? Epitopes which are heat and acid resistant are more likely to have clinical relevance. The major allergen present in codfish, Gad c1, demonstrates remarkable heat and acid resistance. Individuals should be advised to avoid cross-reacting species of fish. The cross-reactivity *in vitro* that is evident between grass pollen and other food cereals such as wheat flour does not seem to be clinically relevant.[30]

3.7.3 The measurement of total IgE

Overall, mainly because of the wide range of total IgE in the population, total IgE is not a very useful diagnostic test to identify allergic individuals. In adults, raised IgE levels carry a high predictive value of IgE-mediated disease, but normal values do not rule it out. In children, total IgE may be more able to distinguish the allergic individual from the non-allergic one. Kjellman found in children that it was both reflective of risk of atopy, correlating well with biparental history of atopic disorders, but also predictive in asymptomatic children of future atopic illness. Children and adults with sensitivities to many allergens, and with multi-end organ involvement, were more likely to have a raised total IgE.[31]

3.7.4 Basophil/leucocyte histamine release test

This test assesses the presence of cell-bound specific IgE. It is based upon the measurement of histamine released from antigen-challenged suspensions of leucocytes. Histamine is released from basophils (a type of leucocyte/white blood cell) as a result of the interaction between allergen and cell-bound specific IgE. Histamine can then be isolated with butanol, then acid, and then the concentration can be assessed spectrofluorometrically or by radio-immunoassay. This concentration of histamine is expressed as a percentage of total cellular histamine. This total cellular histamine is assayed following the lysis of a similar number of non-challenged leucocytes with perchloric acid. It may not directly correlate with specific IgE, as other non-IgE mechanisms may cause the degranulation of basophils. There are several drawbacks to this system. The cells need to be extremely fresh, less than 24 hours old, and only a limited number of allergen challenges can be performed on each aliquot of blood. Fifteen per cent of individuals have leucocytes that do not release histamine after undergoing *in vitro* allergen challenge. There appears to be considerable inter-laboratory variability in the measurement of histamine.[32] Advantages include the opportunity to use fresh allergens, rather than processed and potentially altered allergens used in other tests measuring specific IgE. Bindslev-Jensen and Poulsen calculated the clinical sensitivities and specificities of this test in the adults in their one centre.[30] They estimated a sensitivity as high as 100% when fresh milk was used as the challenge allergen with a specificity of 87%. The least sensitive measure seems to be as low as 50% with commercially prepared egg and milk. These commercially prepared allergens have specificities of 67% for egg and 100% for milk. This test certainly provides an alternative to immunoassays of specific IgE, in particular for rare food allergens.

3.7.5 Other non-IgE antibodies

The body also is capable of producing other types of antibodies such as IgM, IgG and IgA against foods. Some studies have claimed a role for IgA-secreting cells, which have been shown to rise after ingestion of a particular foodstuff, or IgG4 that is said to correlate with clinical hypersensitivity. No studies have been able to demonstrate the role of these antibodies in the pathophysiology of food allergy. Food-specific non-IgE antibodies seem to be much more likely to reflect the particular diet of the individual, a normal phenomenon rather than diagnostic of disease.

The diagnosis of gluten-sensitive enteropathy (coeliac disease) is an exception. The detection of specific non-IgE antibodies aids the diagnosis of coeliac disease. Gastrointestinal auto-antibodies of the IgA class (anti-reticulin and anti-endomyseal antibodies) are raised in this condition. They are classed weakly, moderately and strongly positive. Moderately and strongly positive levels of the antibodies are both sensitive and specific when assessed against challenges of gluten coupled with intestinal biopsy (see later).[6]

3.7.6 Specific T cells: lymphocyte stimulation tests

An increase in peripheral blood mononuclear cells has been demonstrated *in vitro* in people with atopic dermatitis with a late response to milk. T cells expressing cutaneous lymphocyte antigen have been demonstrated in children with milk-dependent eczema. The clinical relevance of these findings remains unclear, and they are yet to be applied in the clinical setting.

3.7.7 Plasma histamine and tryptase

Histamine has been shown to rise following a positive food challenge. It can be measured *in vitro* with difficulty owing to its short half-life. It has about 10% false positives and can only really add to the clinical impression following a DBPCFC.[30] Tryptase is confined to the mast cell. It does not rise reliably following food challenge, nor reliably in many fatal or near-fatal cases of food allergy.[30] Neither has diagnostic value in the routine setting.

3.7.8 Other cell mediators

Monitoring of eosinophils and their products may prove to be a useful outcome measure when assessing oral food challenges. IL-2 and gamma-interferon have also been reported as increasing after a positive challenge. These await further investigation before assessing their clinical significance.

The measurement of components of the complement cascade does not seem to be useful for the clinical evaluation of a patient with suspected food hypersensitivity, nor does the presence of immune complexes containing IgE or IgG in the serum of individuals following challenge seem to be useful in the diagnosis of food hypersensitivity.[30]

3.8 Other useful tests

3.8.1 Permeability tests

One can measure permeability of the gastrointestinal tract to probes of various sizes or to ratios of sugars. There seems to be a wide spread of permeability in those with food sensitivities such that it virtually completely overlaps with that of healthy patients. Changes in permeability have consistently been shown to occur following positive challenges and it is suggested that they may prove useful in the evaluation of drugs suitable for the treatment of food hypersensitivities.[30]

3.8.2 Gastrointestinal biopsy

Gluten-sensitive enteropathy (coeliac disease) should be diagnosed with serial small bowel biopsies (duodenum or jejunum), demonstrating typical histopathological abnormalities on a normal diet and resolution following a gluten-free diet.[6] It has been argued that as this diagnosis has such significant lifelong dietary implications, it should then be followed by a further biopsy demonstrating relapse upon a gluten challenge. Few clinicians carry out this third biopsy.

Histopathological evidence may be useful in other enteropathies. They give evidence of end-organ disease, but not of the culpable allergen.

3.9 Unproven and inappropriately applied tests

There are both unconventional theories and unproved methods being used within the field of allergy. Critics distinguish between diagnostic methods that remain unproven, those that are experimental and those that are accepted by the peer review process and are regarded as standard practice.

Some diagnostic procedures are particularly associated with certain conditions and certain methods of practising. One such group of clinicians are those who work within the field of clinical ecology, involved in the diagnosis of idiopathic environmental intolerances. There are also alternative or complementary practitioners who diagnose food allergy or intolerance in a variety of ways. A thorough review of the most prominent of these practices has been published by The Royal College of Physicians.[26]

3.9.1 Provocation–neutralisation test

This method is used particularly by clinical ecologists in the diagnosis of idiopathic environmental intolerances, also referred to as allergic toxaemia or tension fatigue syndrome. This diagnosis is given to people with a wide range of symptoms involving many areas and systems of the body. The 'causes' listed are many, ranging from reactions to synthetic products, naturally occurring foods, viruses, fungi and even some endogenous hormones such as progesterone. There is no evidence of inflammation or organ dysfunction to support the causal model of immunological hypersensitivity. None of the other non-immunological models are supported by experimental or controlled clinical studies. None of the various laboratory tests used to support the diagnosis have been found to be consistently abnormal.

The provocation–neutralisation test is the main tool used in this area. It has been criticised for the lack of a standard protocol. It involves the exposure of patients to fivefold dilutions of subcutaneously or intracutaneously injected allergen or chemical extract. Occasionally these extracts are given sublingually. For the 10 minutes after each injection the patient records their symptoms. Serially higher doses are administered until the appearance of symptoms is elicited. A progressive series of lower concentrations are then administered until a dose is reached at which the patient experiences no symptoms. This is termed the 'neutralising dose' and is used to determine future treatment doses.

The main criticisms of this test are the lack of theoretical background to support it as a diagnostic procedure and the inconsistency of the pathophysiological explanations with an understanding of immunology. There is also little allowance for spontaneously occurring symptoms and no negative controls. The issue of safety is not addressed. There is a potential danger of anaphylaxis when applying sublingual preparations to patients with a known or unknown potentially serious IgE sensitivity to an allergen.

3.9.2 Cytotoxic test

This is an *in vitro* test based upon particular morphological changes in unstained whole blood cells when challenged with dried food extracts. The changes are viewed microscopically. There is no scientific support either pathophysiologically or clinically.

3.9.3 Electro-dermal testing/electro-acupuncture

This is a test used by alternative or complementary health workers. The device used for this *in vivo* test is made up of a galvanometer that measures the activity of the skin at designated acupuncture points. The patient holds the negative electrode in one hand, while the positive electrode is pressed upon the points. Vials of food extracts in contact with an aluminium plate are also within the circuit. A drop in electrical current is diagnostic of an allergy to that particular food. There is no clearly described theory behind the procedure, and furthermore no clinical or scientific evidence that electro-dermal testing can diagnose food allergy.

3.9.4 Applied kinesiology

Practitioners of this technique test the strength of various groups of the patient's muscles in a subjective manner, while the patient is holding containers of the allergen. Again there is no scientific or clinical evidence to support its use.

3.9.5 Inappropriately applied clinical tests

These include clinical tests that are in common use but have not been shown to identify with any specificity or sensitivity food allergy. An example is alterations in a patient's pulse rate being indicative of a positive immunologically mediated food reaction.

3.9.6 Inappropriately applied laboratory-based tests

There are laboratory-based tests, some in common usage, which have not been shown to identify food allergies or intolerances with any accuracy. Lymphocyte subset counts and lymphocyte function assays are useful for diagnosing congenital or acquired lymphocyte cellular immunodeficiency states, but not allergic disease. Cytokines and their receptors are involved at many levels of the immune response. The correlation of assays with disease, and in particular their diagnostic value, is yet to be established.

It has been suggested that food immune complexes may play a role in those who claim a delayed (>2 hours) adverse response to foods. These are able to be detected using solid phase radioimmunoassay. Even within autoimmune diseases that rely on their presence as diagnostic, their role is unclear. Their clinical relevance is doubtful for two further reasons. They have not been subjected to comparison with DBPCFC, enabling the calculation of sensitivities and specificities. It is also thought that they are probably a normal phenomenon.

Laboratories are able to measure very small amounts of chemicals present in samples of biological material. These are certainly very important to measure where there is clinical concern over toxicity, as with serum levels of lead in cases of concern over lead poisoning, for example. There is no evidence that the chemicals implicated have any part to play in allergic disease. There is also no evidence that deficiencies in vitamins, minerals and amino acids play any part in the allergic response, and thus measuring them is unlikely to be of diagnostic help.

3.10 Summary

Any diagnostic tests only supplement the impression or information already gained from history taking and examination.

The most well-defined mechanism of food allergy is that due to a Type I immunological reaction, mediated by the specific antibodies, IgE. There are other mechanisms of food allergy or intolerance that are much less well defined.

Most well-validated tests have been developed to identify the presence of specific IgE to food allergens either *in vivo* or *in vitro*. Even with these well-validated tests there are still problems in particular with their specificity.

There are diagnostic tests that are unproven, diagnostic tests that are experimental and have yet to move beyond the research arena, and tests which, though valid, have no place in the diagnosis of food intolerance and allergies.

Double-blind placebo-controlled food challenges remain the 'gold standard' test in use when investigating food allergies and intolerances.

3.11 References

1 ISOLAURI E and TURJANMAA K, 'Combined skin prick and patch testing enhances identification of food allergy in infants with atopic dermatitis', *J Allergy Clin Immunol*, 1996 **97** (1) 9–15.

2 SACKETT D L, RICHARDSON W S, ROSENBERG W and HAYNES R B, *Evidence-based Medicine*, Churchill Livingstone, New York, 1997.

3 METCALFE D D, 'Food allergy in adults'. In Metcalfe D D, Sampson H A and Simon R A (eds), *Food Allergy: Adverse Reactions to Foods and Food Additives*, Blackwell Science, USA, 1997.

4 CHAMPION R, ROBERTS S, CARPENTER R and ROGER J, 'Urticaria and angioedema; a review of 554 patients', *British Journal of Dermatology*, 1996 **81** 588–97.

5 LAKE A M, 'Food protein-induced colitis and gastroenteropathy in infants and children'. In Metcalfe D D, Sampson H A and Simon R A (eds), *Food Allergy: Adverse Reactions to Foods and Food Additives*, Blackwell Science, USA, 1997.

6 FEIGHERY C, 'Coeliac disease: fortnightly review', *BMJ,* 1999 **319** 236–9.
7 WEBER R W and VAUGHAN T R, 'Neurological reactions to foods and food additives'. In Metcalfe D D, Sampson H A and Simon R A (eds), *Food Allergy: Adverse Reactions to Foods and Food Additives*, Blackwell Science, USA, 1997.
8 COOKE W J and HOLMES G K T, 'Coeliac disease, inflammation, bowel disease and food intolerance'. In Lessof M H (ed.), *Clinical Reactions to Food*, John Wiley, Chichester, 1983.
9 DENMAN A M 'Allergy and joint complaints'. In Lessof M H, Lee T H and Kenemy D M (eds), *Allergy: an International Textbook*, John Wiley, Chichester, 1987.
10 RIX K J B, PEARSON D J and BENTLEY S J, 'A psychiatric study of patients with supposed food allergy', *Br J Psychiatry*, 1984 **145** 121–6.
11 DAVID T, 'The overworked or fraudulent diagnosis of food allergy and food intolerance in children', *J Roy Soc Med*, 1985 **78**(S 5) 21–31.
12 NETHERCOTT J R and COOLEY J E, 'Getting the most out of patch testing', *Curr Opinion Dermat*, 1995 **2** 10–17.
13 SALVAGGIO J E, 'Diagnosis and management of hypersensitivity pneumonitis', *Hosp Prac*, 1980 **Nov** 93–103.
14 SALVAGGIO J E and O'NEILL C E, 'Occupational asthma caused by organic dusts and chemicals', *Proc XII Int Congr on Allergology and Clinical Immunology*, 486–90, St Louis, C V Mosby, 1986.
15 ZETTERSTROM O, OSTERMAN K, MACHADO L and JOHANSON S G O, 'Another smoking hazard: raised serum IgE concentration and increased risk of occupational allergy', *BMJ*, 1981 **283** 1–7.
16 RISCH N, 'Linkage strategies for genetically complex traits. II The power of affected relative pairs', *Am J Hum Genet*, 1990 **46** 229–41.
17 HANIFIN J M and RAJKA G, 'Diagnostic features of atopic dermatitis', *Acta Derm Venereol Suppl (Stockh)*, 1980 **92** 44–7.
18 BUCCAFOGLI A, VICENTINI L, CAMERANI A, CIGLIATI P, D'AMBROSI A and SCHOLOZZI R, 'Adverse food reactions in patients with grass pollen allergic respiratory disease', *Ann Allergy*, 1994 **73** 301–8.
19 SAMPSON H A, MENDELSON L M and ROSEN J P, 'Fatal and near-fatal anaphylactic reactions to food in children and adolescents', *N Engl J Med*, 1992 **327** 380–4.
20 SAMPSON H A and ALBERGO R, 'Comparison of results of skin tests, RAST, and double-blind, placebo-controlled food challenges in children with atopic dermatitis', *J Allergy Clin Immunol*, 1984 **74** 26–33.
21 BOCK S, SAMPSON H A, ATKINS F M, ZEIGER R S, LEHRER S, SACHS M, BUSH R K and METCALFE D D, 'Double-blind, placebo-controlled food challenge (DBPCFC) as an office procedure: a manual', *J Allergy Clin Immunol*, 1988 **82**(6) 986.
22 BUTCHER B T, HAMMOND Y Y and HENDRICK D, 'Occupational asthma: identification of the agent'. In Gee J B L (ed.), *Occupational Lung Disease*, Churchill Livingstone, New York, 1984.

23 REID M J, LOCKEY R F and TURKELTAUB P C *et al.*, 'Survey of fatalities from skin testing and immunotherapy 1985–1989', *J Allergy Clin Immunol*, 1993 **92** 6–15.
24 SAMPSON H A, 'Food allergy. Part 2: Diagnosis and management', *J Allergy Clin Immunol,* 1999 **103** 981–9.
25 DREBORG S (ed.), 'Skin tests used in type I allergy testing. Position paper', *Allergy*, 1989 **44** (S 10).
26 ROYAL COLLEGE of PHYSICIANS, *Allergy: Conventional and Alternative Aspects*, RCP, London, 1992.
27 HATTEVIG G, KJELLMAN B and BJORKSTEM B, 'Clinical symptoms and IgE to common food proteins and inhalants in the first seven years of life', *Clin Allergy*, 1987 **17** 571–8.
28 VEIEN N K, 'Systemically induced eczema in adults', *Acta Dermato-Venereologica*, S147 ISSN0365-8341.
29 MERRETT T G, 'Quantification of IgE both as total immunoglobulin and as allergen-specific antibody'. In Kay A B (ed.), *Allergy and Allergic Diseases*, Blackwell Science, 1997.
30 BINDSLEV-JENSEN C and POULSEN L K, 'In vitro diagnostic methods in the evaluation of food hypersensitivity'. In Metcalfe D D, Sampson H A and Simon R A (eds), *Food Allergy: Adverse Reactions to Food and Food Additives*, Blackwell Science, USA, 1997.
31 KJELLMAN N I M, JOHANSSON S G O and ROTH A, 'Serum IgE levels in healthy children quantified by a sandwich technique (PRIST)', *Clin Allergy*, 1976 **6** 51–9.
32 NARGAARD A, SKOV P S and BINDSLEV-JENSEN C, 'Egg and milk allergy in adults: comparison between fresh foods and commercial allergen extracts in skin prick test and histamine release from basophils', *Clin Exp Allergy*, 1992 **22** 940–7.

4

Symptoms of food intolerance

J. O'B. Hourihane, Institute of Child Health and Great Ormond Street Hospitals, London

4.1 Introduction

Food allergy has been defined as 'a reaction to a food, which is both reproducible and associated with evidence of an abnormal immunological reaction to the food (mediated by antibody or T-lymphocytes or both)'. The Food and Agricultural Organisation of the United Nations defined food allergy as 'an untoward reaction [to food] due to an immunological mechanism'.[1] These broad headings distinguish food allergy from food intolerance depending on the immunological basis of a reaction. Food intolerance is defined generally as a reproducible but not immunologically mediated reaction. Food aversion is considered to be a psychologically mediated non-reproducible reaction. The term reproducible is important: many people report allergic or intolerance reactions that, when stringently investigated, are found to be spurious. For a reaction to be considered allergic or intolerant in nature, psychological and emotional factors need to be controlled – by exclusion. This is the basis of the double-blind placebo-controlled food challenge (DBPCFC), where patient and observer bias are eliminated or minimised.[2] Allergic reactions and intolerant reactions are reproducible in a double-blind placebo-controlled food challenge *if the threshold dose is reached.* Food aversion is not reproducible in blinded food challenges. The rest of this discussion is restricted to food allergy and food intolerance reactions.

4.2 How to distinguish intolerance from allergy

Several clues in the history of a reaction will lead a clinician or interested party towards classifying that reaction as either allergic or intolerant. Some of these

Table 4.1 The 'big 8' foods that cause more than 90% of IgE-associated reactions in children

Milk	Tree nuts (and sesame)
Egg	Wheat
Soy	Fish
Peanut	Shellfish

features relate to the foods, but some relate to the subjects reporting the symptoms.

4.2.1 Foods commonly associated with allergy (Table 4.1)

Yunginger *et al.*[3] and Sampson *et al.*[4] showed the most common cause of severe food-related allergic reactions in adults and older children to be peanuts, crustaceans, shellfish, tree nuts and fish. In selected American children with atopic dermatitis (eczema), Burks *et al.*[5] showed that skin prick testing with eight foods identified 99% of subjects who reacted to a food in DBPCFC, even if the food causing the reaction in the challenge had not been one of the foods used for skin testing. Or, put another way, subjects who reacted to an unusual food nearly always had a positive skin prick test (SPT) to one of the eight foods used for screening with or without associated symptoms on exposure to that food. Such studies need to be repeated in different populations of subjects. There are clearly geographical variations regarding these foods because the lists involved in reactions in Britain[6] are like American lists but European studies give slightly different figures regarding allergic reactions to foods. André *et al.*[7] followed 580 French patients with pathological reactions to foods from 1984 to 1992. Wheat (39%), was the most common cause of sensitisation, followed by peanuts (37%), crab (34%), celery (30%) and soy (30%). Novembre *et al.*[8] found that foods caused 54 out of 95 episodes of anaphylaxis in 76 Italian children. Fish was the commonest causative food (16 cases), followed by cow's milk (12), nuts (7) egg (6), fruit (6) and cereals (3).

These foods are associated with the development of antibodies (sensitisation) that typify the immediate allergic response – IgE antibodies. After sensitisation very low doses can cause immediate allergic reactions.[9] For instance, 76% of subjects with peanut allergy experience symptoms within five minutes and 93% within 30 minutes, and 90% reported symptoms after eating less than one peanut.[10]

In contrast, intolerance reactions may require repeated doses, have a characteristically slower onset and resolution and are more commonly associated with reactions to multiple foods.[11]

4.2.2 Characteristics of patients with food intolerance

The most widely quoted study is by Parker *et al.*[12] They were able to divide 45 adults reporting food-related complaints into two groups. The first group of 22 subjects reported reactions to the common allergenic foods – legumes, tree nuts, crustaceans and fish. Twenty-one out of the 22 subjects in this group had positive skin prick tests to the offending food. The second group reported reactions to food such as sugar, wheat, egg, cured meat and yeasts. Only four out of the 23 subjects in group 2 had a positive skin prick test in this group (evidence of IgE) that supported their reported reactions (chi-squared 24.68, p value $<$ 0.0001). The second group's symptoms started at an older age – 28.9 years versus 17.1 years ($p = 0.0015$) – and were related to a much broader range of foods – 25.6 versus an average of 5.5 ($p = 0.0002$).

In this key study Parker *et al.* showed that the group of adults with allergic-type symptoms were significantly more likely to suffer swelling and respiratory symptoms than the group with non-allergic complaints and negative skin prick tests. The group with non-allergy-mediated complaints reported significantly more non-specific problems such as neurological symptoms (headache, fainting, numbness), gastrointestinal symptoms (bloating and distension but not pain, vomiting or diarrhoea) and musculoskeletal symptoms (cramps and stiff joints).

Briefly, intolerance reactions are more common in adults and a wide range of responsible foods and symptoms induced are demonstrated.[11,12] It must be remembered that not all IgE-mediated disease occurs immediately. There is a well-described phenomenon of late-phase IgE reactions with late urticaria (itchy hives) and oedema (swelling) within the first 24 hours of exposure to the foods. It can be difficult to distinguish clinically this 24–48 hour reaction from that which is caused by non-IgE-mediated immunological reactions such as those that cause an exacerbation of eczema. Often the temporal association with a dietary exposure to allergen is the only clue.

4.3 Oral allergy syndrome

The constellation of immediate symptoms less than one hour after exposure and usually confined to the mouth has been called the oral allergy syndrome (OAS), first characterised in 1987 by Amlott *et al.*[13] The initial group of 36 subjects was broadly divided into those whose symptoms did not progress (50%) and those who responded to larger doses of allergen, with more severe reactions. For each individual subject the quantity of food required to cause OAS and other symptoms varied.[13]

The typical symptoms of OAS involve a tingling in the lips, swelling of the tongue and maybe a feeling of swelling in the back of the throat. Patients often will complain of something stuck in their throat, feeling like a cherry stone or a food bolus. It can be difficult to distinguish this from more severe laryngeal oedema (swelling of the voice box), but the latter usually causes noisy inspiration or repetitive coughing. OAS needs to be distinguished from the early warning features of a gradually generalising reaction.

4.3.1 Foods that cause OAS

The common foods that cause OAS in Britain are apples and stone fruits. It is a very characteristic feature of OAS that some people react to foods that are raw but not to foods that are cooked and to the peel rather than the pulp.[14] The classic example is being unable to tolerate raw apples even by touching a fruit but being able to tolerate apple pie.

4.3.2 OAS and pollen allergy

The frequency with which OAS occurs in subjects with pollen allergy is notable. Up to 40% of subjects with birch and ragweed allergy suffer OAS.[15] Ragweed allergy is particularly associated with reactions to bananas and melons, and birch allergy with celeriac, apple and hazelnut allergens. The basis of the latter is thought to be homology between the relevant allergens, particularly Bet V 2 from birch, Mal d 1 from apple, and Bet v1 and Apig 2 from celeriac.[16] Treatment of pollen allergy with immunotherapy has abrogated associated OAS reactions.[17]

4.4 Evolution of allergic reactions

Two of the most important features that distinguish allergic reactions associated with allergen-specific IgE are the rapid onset of symptoms, usually within 5–10 minutes of exposure to foods, and the gradual resolution in the course of one or two hours. Most mild to moderate reactions occur within this time frame. Mild to moderate reactions are generally defined as reactions confined to the skin or gastrointestinal tract, while severe reactions are those that threaten the airway or cause a fall of blood pressure. It can be very difficult in most subjects to predict when a reaction is becoming so severe that treatment must be initiated. Severe reactions can gradually evolve from relatively minor symptoms and can form a second phase of response once the initial symptoms have resolved, or they can gradually develop slowly and persist for considerable periods of time. This variation in the presentation of severe symptoms needs to be specifically sought in the history. Most reactions that are due to intolerance are slow in onset, and are often dose-related or related to the duration of exposure, such as from eating a food on several occasions over several days.[11,12] In contrast, allergic reactions are usually due to small doses and start immediately upon exposure.[9,10]

4.5 Clinical categorisation of allergic reactions

In a series of 62 adults and children with peanut allergy, Ewan[18] divided patients into those whose separate symptoms were:

- skin symptoms only
- symptoms involving skin and airway

- significant fall in blood pressure or loss of consciousness.

Out of the 62 patients, 20 had skin changes only, 33 had evidence of airway involvement with laryngeal oedema or wheezing, and nine had evidence of a significant fall in blood pressure.[18] The categorisation of laryngeal oedema is discussed below. Contact symptoms are common in food-related allergic diseases, especially in children and those with irritated or inflamed skin diseases such as eczema. These symptoms are very rare in people with food intolerance, and most adults (99%) with the syndrome of chronic urticaria (bouts of intermittent episodes of itchy hives and swelling that last longer than 6 weeks) do not have food allergy.[19]

Sicherer *et al.*[20] showed that in 102 individuals with peanut allergy, the first reaction is characterised by isolated skin reaction in 49%, by respiratory reaction only in 2%, by both skin and respiratory in 17%, by both skin and gastrointestinal in 7%, and by all three systems in 21%. None of this group of peanut-allergic individuals suffered a significant fall in blood pressure or loss of consciousness on first exposure.[20]

In general, most skin reactions start soon after exposure and are past the worst at about 30 minutes. Occasionally angioedema (swelling) can persist for up to 48 hours. Skin reactions can evolve into mild wheezing (particularly in asthmatics) and may develop into anaphylaxis, although that is unusual. The most worrying scenario is the onset of severe symptoms – a fall in blood pressure or severe wheezing – without skin manifestations that might act as early warning signs. The biphasic nature of some reactions is particularly worrying and can be associated with a fatal outcome.[4] That is why people who have suffered a severe allergic reaction to foods must be observed adequately in hospital, even if overnight admission is required.

4.5.1 Frequency of individual symptoms

Table 4.2 shows the frequencies of each allergic symptom in the first and most recent reactions to peanut in the author's Southampton study.[21] Most symptoms occurred in varying combinations. The most common features were facial swelling and rash and these nearly always occurred together.

Vomiting and abdominal pain (not shown in the table) were rarely isolated features of a reaction. Abdominal pain correlated more strongly with severe symptoms than did vomiting. Abdominal pain also correlated more strongly with severe symptoms than did more minor symptoms such as rash or itch. Abdominal pain may be an under-appreciated symptom of at least moderately severe allergic reactions to peanut.

The most severe symptom, collapse, never occurred in isolation and was always associated with some other preceding symptom. In subjects whose most recent reaction was characterised by collapse (46 subjects) facial swelling was present in 83%, wheeze in 74% of cases, itch in 67%, rash in 59% and vomiting in 50% of cases.[10,22]

Table 4.2 Features of first and most recent reactions to reaction to peanut

Symptom	First reaction Number (%)	Most recent reaction Number (%)
Facial swelling	444 (72)	360 (69)
Rash	399 (64)	308 (59)
Itch	342 (55)	316 (61)
Vomiting	240 (39)	209 (40)
Wheeze	239 (38)	246 (47)
Breathing difficulty	229 (37)	231 (44)
Cyanosis	88 (14)	53 (10)
Collapse	42 (7)	44 (8.5)

4.5.2 Does severity always increase with successive reactions?

It might appear from Table 4.2 that symptoms are constant from one reaction to the next. In general it is true that subjects will have similar reactions if exposed to similar doses in similar situations. It is clear, however, that doses and circumstances vary considerably and therefore reactions can vary considerably too. This is reflected in Table 4.3, which shows how symptoms in the author's Southampton study[21] had changed in individuals between the first reaction and the most recent reaction, according to physician review of the reported symptoms. Less than half (41%) of subjects with mild first reactions did not suffer a worse reaction on the most recent exposure. The mild reactors who progressed to more severe reactions were divided almost equally between moderate and severe reactions on the most recent reaction, and 86% of subjects with moderate first reactions suffered a similar or worse reaction on most recent exposure to peanut. Reportedly severe first reactions were predictably stable with 78% suffering a further severe reaction on most recent reaction. Only 9% of severe first reactions were followed by a mild reaction and 13% by a moderate reaction.[21]

Table 4.3 Comparison of first and most recent reactions to peanut

First reaction	Most recent reaction			
	Mild	Moderate	Severe	Total
Mild	44	29	34	107
Moderate	27	111	59	197
Severe	21	29	173	223
Total	92	169	266	527

4.6 Anaphylaxis

4.6.1 Definition

There is a difference in implication between British and American usage of the term anaphylaxis. As used by Americans, anaphylaxis usually relates to all generalised IgE-mediated reactions that are characterised by the above symptoms, irrespective of severity. In the case of the 'big 8' foods that cause most cases of food-related allergic reactions, it may act as a warning of the potential for anaphylaxis, even if it has not happened yet. The British use of the term is reserved for those in whom exposure to an allergen has caused a significant fall in blood pressure or severe wheezing.

4.6.2 How common is anaphylaxis?

Much local data has been collated in order to try to gain estimates of how common anaphylaxis might be. Yocum and Khan[22] estimated that food caused one-third of anaphylactic reactions that occurred outside hospitals in the US. Danish data suggest prevalence of one in 30 000 population per year.[23]

A prospective study by Bock in Denver, Colorado, found 25 cases of food-related anaphylaxis over a three-year period, a third of which were due to peanut.[24] This number of reactions in a state whose population was 3.3 million gives a minimum incidence of severe reactions to foods of approximately one per 264 000 people per annum. Extrapolation of this minimum figure to the entire US population would suggest a national incidence of 950 individuals per annum.

A similar study in Munich, over the course of one calendar year (1992), showed that food caused 17 of 150 cases (11.3%) of anaphylaxis requiring emergency room treatment. The minimum overall incidence of food-related anaphylaxis in Munich was estimated to be 1.1 per 100 000 people.[25] The incidence of actual reactions would be higher, due to the recalcitrant nature of severe reactions[3,4] and, in the case of peanut, at least, to the difficulty in avoiding exposure[26] and the underreporting of reactions, even to airline staff during a flight,[27] let alone seeking medical help.

A French study[28] supports Bock's findings from Colorado. This multi-centre study investigated the presentation rate of food-induced anaphylactic shock to 46 emergency departments, 29 dermatology units and 19 internal medicine departments. In 794 reported cases of anaphylaxis, food was implicated in 81 cases (10%). Unusually, only 19 patients (23.4%) had known food allergy. The presence of the causative allergen in 'hidden form' contributed to 25 cases (31%) of food-related anaphylaxis. An enhancing factor, such as alcohol consumption or exercise,[29] was present in 221 cases (27.8%).

A retrospective British study suggests that the incidence of food-related anaphylaxis is broadly similar in Britain and continental Europe. Stewart and Ewan[30] analysed clinical records of attendances at the casualty unit of Addenbrooke's Hospital, Cambridge (catchment area population 350 000), over the full calendar year of 1993 and over a three-month period in 1994. They found nine cases of collapse (severe anaphylaxis) and 15 cases of generalised allergic reaction (without hypotension) in 55 000 attendances in 1993. The rate of generalised allergic or severe anaphylaxis (combined total 24 cases) was, therefore, one case in every 2300 casualty attendances, or 6.8 per 100 000 of the local population per year. Three of the nine severe anaphylaxis reactions were due to foods (one per 18 000 casualty attendances, or 1.16 per 100 000 population per year). This figure is very similar to those of Bresser *et al.*[25]

Stewart and Ewan noted an increased rate of diagnosis of anaphylaxis from any cause in the follow-up three-month study the following year, probably due to an increased rate of ascertainment and to the study falling in the peak season for wasp and bee stings. Stings accounted for eight of the nine anaphylactic reactions noted over the three-month period. With increased staff awareness of anaphylaxis the incidence increased to one per 1500 casualty attendances.

4.6.3 Predisposing factors for anaphylaxis

It is clearly established now that pre-existing asthma may exacerbate or predispose to anaphylaxis. This is clearly in keeping with the concept of the target organ of the reaction being important. Sampson's group of fatal and non-fatal reactors were clearly distinguished by the presence of a diagnosis of asthma, particularly if the asthma was poorly controlled. Our study in Southampton has supported this finding (Table 4.4).[10,21]

Other factors that may predispose to severe disease are the use of medications that may interfere with the normal physiological response to an allergen, i.e. thc epinephrine (formerly adrenaline) response. The typical drugs implicated are the cardiac drugs called angiotensin converting enzyme (ACE) inhibitors which inhibit the physiological angiotensin reaction to hypertension, and the more familiar beta blockers prescribed for ischaemic heart disease. The latter may inhibit the generation of a faster heartbeat to maintain organ profusion in the face of allergen-mediated fall in blood pressure.

Alcohol is a clear example of a drug that may work in several ways. In amounts that are consumed socially, it impairs judgement and encourages risk taking and may therefore impair the intellectual response to being exposed to a food known to potentially contain an allergen that is unsafe for the subject (the classic example being the Indian takeaway). Alcohol also works physiologically by causing relaxation of blood vessels, leading to a fall in blood pressure. A fall in blood pressure may be more precipitous if an allergen is consumed during a meal in which alcohol is also consumed.

4.6.4 Is laryngeal oedema a moderate or severe symptom?

The larynx is the narrowest part of the upper airway. Clearly, therefore, any change in the diameter of the larynx (by oedema) is going to have a significant impact on airflow and tissue oxygenation. The distinction between laryngeal

Table 4.4 Association of severity of first reaction to peanut and presence of asthma

Reaction	Asthma present	Asthma absent	Total
Mild	66	51	117
Moderate	121	110	231
Severe	189	80	269
Total	376	241	617

oedema and swelling above the larynx in the back of the throat can be very difficult. In general, laryngeal oedema is characterised by noisy breathing, a change in the character of the voice, and a cough rather than a mere feeling of something stuck in the throat.

Some authors classify laryngeal oedema as a severe symptom due to its categorisation as an airway symptom.[18] However, it is not a common autopsy finding in fatal cases (Sampson, personal communication) and is usually easily treated with epinephrine. It is possible that acute severe laryngeal oedema could cause fatal reactions, although I am not aware of such cases.

4.6.5 Exercise-induced anaphylaxis

Exercise-induced anaphylaxis is an increasingly recognised problem over the past decade or so.[29,31] This may be due to the increased participation in prolonged exercise such as jogging and marathons by the general public. Reactions are usually associated with specific foods, and the onset of anaphylaxis is associated with exercise within two to four hours of eating the food.

It can take considerable detective work on behalf of the patient and the doctor to make the association with particular foods. If a specific food is involved these individuals usually have positive skin prick tests. Foods that are particularly associated are fish, shellfish and wheat. The simple precaution of not eating such a food when exercise is anticipated is usually adequate, although epinephrine may need to be available for these individuals. Avoidance of exercise during the pollen season is also effective.[31]

4.6.6 How long should subjects wait before treating themselves with epinephrine?

The difficulty for subjects in deciding when to treat themselves is one of the most thorny issues of food anaphylaxis. General advice is that children and adults should treat themselves with antihistamines immediately in the form of liquid antihistamines, and that epinephrine should be reserved for those cases in which the reaction is progressing or if, in the rare cases, symptoms are immediately severe without other symptoms. Everybody who has had a severe reaction to peanuts and those who are at risk of reactions should treat themselves appropriately with epinephrine if the first symptoms are progressing. Delay in administration of epinephrine has been identified as contributing to fatal reactions.[4]

4.7 Other symptoms of food-related disease

4.7.1 Eczema

Eczema can be part of a late phase IgE-mediated reaction or a delayed immune reaction to allergen, not mediated by IgE. Eczema is a common feature in people who do not have positive skin prick tests or IgE tests to the allergen. It is therefore only on the basis of a clinical improvement on exclusion of the food and relapse on reintroduction that the diagnosis can be made. Usually the only clinically useful test is an exclusion diet. Patch testing is being investigated as a diagnostic tool for food allergy, particularly in children.[32]

4.7.2 Neurological and musculoskeletal disease

Migraine is a very difficult symptom to describe and has many causes. It is familial and many people describe migraine that is related to foods. Whether it is due to direct effects of molecules that are in the food or an immunological reaction to the food is difficult to establish. The same is true for arthralgia. I am not aware of any double-blind studies that have shown an association between arthralgia and food. The association of foods with symptoms such as ME (myalgic encephalomyelitis) is very difficult to prove or disprove.

4.7.3 Gastrointestinal reactions

Gastrointestinal reactions to foods are very common and not all are associated with positive results on standard tests. A severe feature of food allergy in childhood is called eosinophilic enterocolitis, where the lining of the bowel is filled with cells called eosinophils, which are major factors in local allergic inflammation and reactivity. Patients, usually less than two years old, have severe abdominal pain and bloody diarrhoea, usually made worse by several foods. The treatment of this is the same as for food-related anaphylaxis, i.e. allergen exclusion. In many cases multiple allergens may need to be excluded, and enterocolitis can be a very difficult problem to treat.[33]

Diarrhoeal reactions to foods are common. Isolated gastrointestinal symptoms are a rare manifestation of allergy[20,21] but are a common feature of intolerance reactions. Wheat is a common cause of diarrhoeal reactions and benign wheat intolerance must be distinguished from coeliac disease. In coeliac disease a different class of antibodies (IgA, not IgE) is generated.

4.8 Summary

The presentation of individuals with food-related symptoms appears to be increasingly prevalent. Not all symptoms are associated with defined immunological mechanisms and therefore the range of foods in which intolerance can be described is potentially infinite. A small subset of foods usually accounts for more than 90% of severe reactions to foods that are mediated by or associated with IgE.

Not all individuals with evidence of IgE to foods are at risk of anaphylaxis. Not every individual requires prescription of epinephrine. The subjects who need epinephrine are those who have had severe reactions previously or who have a previous history of asthma. This is, however, the majority of patients and therefore most patients with food-related allergies should have epinephrine available. This does not mean, however, that administration of epinephrine should be an automatic step in every case. Most reactions are to small doses and either are self-limiting or respond well to oral antihistamines.

The warning signs of reactions and preparedness to treat despite uncertainty as to how the reaction will progress are the cornerstones of management for these patients. These people need to be aggressively polite in enquiring about the contents of the meals that they are being served by persons who are not friends or family and who may not necessarily have a vested interest in their well-being.

4.9 Sources of further information and advice

The Ministry of Agriculture, Fisheries and Food has produced information packs for the catering industries on how to anticipate the problems that might arise from allergic individuals in their restaurants. The Anaphylaxis Campaign, the British Allergy Foundation and their American counterpart, the Food Allergy Network, are excellent sources of rationally prepared and non-hysterical advice for individuals affected by food allergy and intolerance. The American Academy of Allergy, Asthma and Immunology has a website that has free patient information available, links to other sites and access to American physicians who may treat individuals. The Anaphylaxis Campaign has recently gone online. The British Allergy Foundation website is rather rudimentary at present.

Websites and addresses

American Academy of Allergy, Asthma and Immunology: www.aaaai.org

Anaphylaxis Campaign
PO Box 149, Fleet, Hampshire GU13 0FA
Tel. 01252 542029
www.anaphylaxis.org.uk

British Allergy Foundation
Deepdene House, 30 Bellegrove Road, Welling, Kent DA16 3PY
Tel. 020 8303 8525
Fax. 020 8303 8792
www.allergyfoundation.com

Food Allergy Network: www.foodallergy.org

Ministry of Agriculture, Fisheries and Food: www.maff.gov.uk

Recommended reading

Ortolani C, Bruinzeel-koomen C, Bengtsson C *et al.* 'Controversial aspects of adverse reactions to food', *Allergy*, 1999 **54** 27–45.

Sampson H A, 'Food allergy. Part 2: Diagnosis and management', *J Allergy Clin Immunol*, 1999 **103** 981–9.

4.10 References

1 UNITED NATIONS FOOD AND AGRICULTURE ORGANISATION, *Report of the FAO Technical Consultation on Food Allergies*, Rome, 1995.

2 BOCK S, SAMPSON H A, ATKINS F M, ZEIGER R S, LEHRER S, SACHS M, BUSH R K and METCALFE D D, 'Double-blind, placebo-controlled food challenge (DBPCFC) as an office procedure: a manual', *J Allergy Clin Immunol*, 1988 **82**(6) 986–97.

3 YUNGINGER J W, SWEENEY K G, STURNER W Q, GIANNANDREA L A, TEIGLAND J D, BRAY M, BENSON P A, YORK J A, BIEDRZYCKI L, SQUILLACE D L *et al.*, 'Fatal food-induced anaphylaxis', *JAMA*, 1988 **260**(10) 1450–2.

4 SAMPSON H A, MENDELSON L, and ROSEN J P, 'Fatal and near-fatal anaphylactic reactions to food in children and adolescents', *N Engl J Med*, 1992 **327**(6) 380–4.

5 BURKS A W, JAMES J M, HIEGEL A, WILSON G, WHEELER J G, JONES S M and ZUERLEIN N, 'Atopic dermatitis and food hypersensitivity reactions', *Pediatrics*, 1998 **132** 132–6.

6 PUMPHREY R S and STANWORTH S J, 'The clinical spectrum of anaphylaxis in north-west England', *Clin Exp Allergy*, 1996 **26**(12) 1364–70.

7 ANDRÉ F, ANDRÉ C, COLIN L, CACARAU F and CAVAGNA S, 'Role of food allergens and of allergen consumption in the increased incidence of food sensitisation in France', *Toxicology*, 1994 **93**(i) 77–83.

8 NOVEMBRE E, CIANFERONI A, BERNARDINI R, MUGNAINI L, CAFFARELLI C, CAVAGNI G, GIOVANE A and VIERUCCI A, 'Anaphylaxis in children: clinical and allergologic features', *Pediatrics*, 1998 **101**(4) e8.

9 HOURIHANE J O'B, KILBURN S A, NORDLEE J A, HEFLE S L, TAYLOR S L and WARNER J O, 'An evaluation of the sensitivity of subjects with peanut allergy to very low doses of peanut protein: a randomized, double-blind,

placebo-controlled food challenge study', *J Allergy Clin Immunol*, 1997 **100**(5) 596–600.

10 HOURIHANE J O'B, KILBURN S A, DEAN T P and WARNER J O, 'Clinical characteristics of peanut allergy', *Clin Exp All*, 1997 **27** 634–9.

11 PARKER S L, KRONDL M and COLEMAN P, 'Foods perceived by adults as causing adverse reactions', *J Am Diet Assoc*, 1993 **93**(1) 40–4.

12 PARKER S L, LEZNOFF A, SUSSMAN G L, TARLO S M and KRONDL M, 'Characteristics of patients with food-related complaints', *J Allergy Clin Immunol*, 1990 **86**(4 Pt 1) 503–11.

13 AMLOT P L, KEMENY D M, ZACHARY C, PARKES P and LESSOF M H, 'Oral allergy syndrome (OAS): symptoms of IgE-mediated hypersensitivity to foods', *Clin-Allergy*, 1987 **17**(1) 33–42.

14 FERNANDEZ-RIVAS M and CUEVAS M 'Peels of rosaciae fruits have a higher allergenicity than pulps', *Clin Exp Allergy*, 1999 **29** 1239–47.

15 BIRCHER A J, VAN MELLE G, HALLER E, CURTY B and FREI P C, 'IgE to food allergens are highly prevalent in patients allergic to pollens, with and without symptoms of food allergy', *Clin Exp Allergy*, 1994 **24**(4) 367–74.

16 EBNER C, HIRSCHWEHR R, BAUER L, BREITENEDER H, VALENTA R, EBNER H, KRAFT D and SCHEINER O 'Identification of allergens in fruits and vegetables: IgE cross-reactivities with the important birch pollen allergens Bet v 1 and Bet v 2 (birch profilin)', *J Allergy Clin Immunol*, 1995, **95** (5 Pt 1) 962–9.

17 KELSO J M, JONES R T, TELLEZ R and YUNGINGER J W, 'Oral allergy syndrome successfully treated with pollen immunotherapy', *Ann Allergy Asthma Immunol* 1995 **74**(5) 391–6.

18 EWAN P W, 'Clinical study of peanut and nut allergy in 62 consecutive patients: new features and associations', *BMJ*, 1996 **312** 1074–8.

19 CHAMPION R, ROBERTS S, CARPENTER R and ROGER J, 'Urticaria and angioedema; a review of 554 patients', *British Journal of Dermatology*, 1969 **81** 588–97.

20 SICHERER S H, BURKS A W and SAMPSON H A, 'Clinical features of acute allergic reactions to peanut and tree nuts in children', *Pediatrics*, 1998 102(1):e6.

21 HOURIHANE J O'B, *Clinical and immunological features of peanut allergy*, Thesis, University of Southampton, 1996.

22 YOCUM M W and KHAN D A, 'Assessment of patients who have experienced anaphylaxis: a 3-year survey', *Mayo Clinic Proceedings*, 1994 **69**(1) 16–23.

23 SORENSEN H T, NIELSEN B and OSTERGAARD-NIELSEN J, 'Anaphylactic shock occurring outside hospitals', *Allergy*, 1989 **44**(4) 288–90.

24 BOCK S A, 'The incidence of severe adverse reactions to food in Colorado', *J Allergy Clin Immunol*, 1992 **90** (4 Pt 1) 683–5.

25 BRESSER H, SANDNER C and RAKOSKI J, 'Anaphylactic emergencies in Munich in 1992 (Abstract)', *J Allergy Clin Immunol*, 1995 **95** (1 Pt 2) 368.

26 BOCK S A and ATKINS F M, 'The natural history of peanut allergy', *J Allergy Clin Immunol*, 1989 **83** (5) 900–4.

27 SICHERER S H, FURLONG T J, DESIMONE J and SAMPSON H A, 'Self-reported allergic reactions to peanut on commercial airliners', *J Allergy Clin Immunol*, 1999 **104**(1) 186–9.
28 MONERET-VAUTRIN D A and KANNY G, 'Food-induced anaphylaxis. A new French multicenter survey', *Ann Gastroenterol Hepatol Paris*, 1995 **31**(4) 256–63.
29 TILLES S, SCHOCKET A and MILGROM H, 'Exercise-induced anaphylaxis related to specific foods', *J Pediatr*, 1995 **127**(4) 587–9.
30 STEWART A G and EWAN P W, 'The incidence, aetiology and management of anaphylaxis presenting to an accident and emergency department', *QJM*, 1996 **89**(11) 859–64.
31 SHADICK N A, LIANG M H, PARTRIDGE A J, BINGHAM C, WRIGHT E, FOSSEL A H and SHEFFER A L, 'The natural history of exercise-induced anaphylaxis: survey results from a 10-year follow-up study', *J Allergy Clin Immunol*, 1999 **104**(1) 123–7.
32 ISOLAURI E and TURJANMAA K, 'Combined skin prick and patch testing enhances identification of food allergy in infants with atopic dermatitis', *J Allergy Clin Immunol*, 1996 **97** (1 Pt 1) 9–15.
33 SAMPSON H A, 'Food allergy. Part 2: Diagnosis and management', *J Allergy Clin Immunol*, 1999 **103** 981–9.

5

The treatment of food intolerance

S. H. Arshad, The David Hide Asthma and Allergy Research Centre, Isle of Wight

5.1 Introduction: the range of treatments

Treatment of food intolerance is primarily by elimination of the food from the diet. This may be a difficult task for a number of reasons. It may not always be possible to identify the foods responsible for the symptoms. Some foods are consumed so frequently that simple elimination without adequate replacement may lead to nutritional deficiencies. Other foods, such as egg and nuts, may be hidden in prepared foods such as cakes and biscuits, and occasional inadvertent exposure may occur.

5.1.1 Avoidance and elimination diets

An assessment should be made of the severity of the reaction, which gives an indication of the strictness of the recommended avoidance regimen. If the symptoms are severe and life threatening, complete elimination of the food is mandatory. Less severe symptoms may allow some degree of flexibility. A detailed explanation is essential for successful avoidance. Egg, milk, soy and nuts may be hidden in other foods and reading the ingredient list is essential. Alternative food should be suggested and it is essential to make sure that the avoidance diet is nutritionally adequate. A complete dietary history is of utmost importance and may uncover important sources of allergenic foods such as milk. It would also allow the caring physician/dietitian to suggest alternative foods with equivalent nutritional value. For example, soya milk or hypoallergenic formulae can be given for cow's milk intolerance during infancy. The services of a qualified dietitian are extremely useful.

5.1.2 Pharmacotherapy

Drug treatment is of value in acute allergic reactions, as unintentional ingestion of the food may occur. The need for pharmacotherapy depends on the sensitivity to the allergenic foods and severity of the resulting symptoms. Those who are highly sensitive, for example, to peanut, may react to minute amounts of peanut proteins hidden in the packaged food or food contaminated by peanut during preparation. Despite taking extreme care, total elimination cannot be guaranteed. This leads to significant morbidity and mortality. Early treatment with adrenaline may be life saving.

Drug treatment for patients with chronic food allergy and intolerance is not rewarding. Several drugs have been tried but the results are disappointing. Patients with multiple food allergy or intolerance are at a particular disadvantage, as avoidance is difficult and may lead to nutritional deficiencies. For this reason the diagnosis of food intolerance should only be made after careful consideration of history, immunological tests and preferably double-blind placebo-controlled food challenges (DBPCFCs). The practice of diagnosing food intolerance by unproven and non-scientific methods should be actively discouraged.

5.2 Avoidance therapy

5.2.1 General principles

Introduction

The diagnosis of food-related symptoms should not be taken lightly, as food avoidance can be difficult, expensive, disruptive and even harmful to the health of the patient, especially in infants and young children[1] (Table 5.1). The increasing complexity of our food intake and a higher proportion of packaged/cooked foods in our diet make the avoidance of a particular food difficult (Table 5.2). The food industry has become increasingly important in the lives of patients with food allergy and intolerance.[2] Those involved in the processing and packaging of foods should be aware of the basic principals of food intolerance and how changes in the food processing/packaging might affect the lives of millions of people with food intolerance[3] (Table 5.3).

The methods used for the diagnosis include a detailed history and immunological tests, followed by challenge procedures. If a definitive diagnosis

Table 5.1 Problems that may be encountered when one or more foods are excluded from the diet

Nutritional effects, especially on the growing child
Limitation of choice of foods
Overprotection of the child
Non-compliance, especially in teenage children
Continued fear/anxiety despite evidence of the food being tolerated
Social isolation

Table 5.2 Foods that commonly cause allergy or intolerance should be clearly labelled when these are used in a packaged food by the food industry

Foods recommended for specific labelling (even small quantities of these foods should be individually listed as ingredients):

- milk and milk products
- egg and egg products
- legumes: peas, peanut, soyabeans and their products
- barley, oat, wheat and rye
- tree nuts: brazil nut, hazelnut, walnut, almond, cashew nut, pecan
- poppy seeds, sesame seeds and their products
- fish and fish products
- crustaceans and other shellfish
- sulphite (>10 mg/kg)

(Report of the FAO Technical Consultation on Food Allergies. Rome, Italy, 13–14 November 1995, Food and Agriculture Organisation of the United Nations, 1995.)

can be made with the help of one or more of these procedures, the advice on elimination of the food is relatively simple. However, this is not always possible, as the history may be inconclusive and immunological tests have a limited value in the identification of the suspected food. Even DBPCFC may not be helpful in some patients. Moreover, DBPCFCs are not practical or feasible in all patients.

Table 5.3 The food industry can play an important role in the management of the patient with intolerance to food

Producers must be aware of the importance of clear labelling and the risk to the allergic individual of even small amounts of hidden foods

Legislation and government control is important for labelling and control of food production

All ingredients, particularly highly allergenic foods, should be clearly stated

Labelling must be readable and consumers must be able to understand the text

Highly allergenic foods such as nuts should not be added to a previously known product without clear warning

Producers should be aware of the changes in allergenicity that might occur during processing of foods, as this can both increase or decrease allergenicity

Only food additives that are necessary for the texture, preservation and nutritional adequacy of a particular food should be allowed. These should be clearly stated among the ingredients

It is necessary to avoid contamination of foods during production and packaging between different foods

Food manufacturers should invest in research in the production of hypoallergenic foods or foods that might be suitable for patients with intolerance to a particular food

Some patients are allergic or intolerant to multiple foods or food additives where the identification may be particularly difficult. A different approach in the management is required in these patients.

Strictness of the diet
Whether food should be completely eliminated from the diet or merely avoided depends on the severity of the symptoms. It is of course more difficult to eliminate a food from the diet completely, as this requires reading labels and being aware of the composition of food bought in the supermarket or consumed in the restaurant.[1] This may not be necessary, as small amounts of hidden food, for example egg in cake and biscuits, may well be tolerated by some patients. Others may react to even small amounts and these patients should eliminate the food completely. Therefore, contamination must be avoided during production, storage, cooking and serving.

Heating of foods
Heating generally reduces the allergenicity of the food antigens.[4] Vegetable proteins such as banana, apple and potatoes are thermolabile, although peanut retains its allergenicity when heated. The ripening of fruit and vegetables such as peas, peanuts, beans and tomatoes increases their allergenicity. Allergens from animal proteins such as milk, egg and fish are relatively thermostable and cooking does not diminish their allergenicity. However, meat albumin is thermolabile. The knowledge of changes in the allergenicity of foods may be useful when advising patients in food allergen avoidance.

Need for review
The intolerance to a food is not always for life, especially in children, where food allergy is common but often transient. Patients may continue to practise avoidance for years when this is no longer necessary. Review of the diet is important at regular intervals. The nature of food and previous reaction dictates the course of action. Peanut and tree nut allergy tends to be lifelong despite complete avoidance. However, reports of patients who have lost peanut and tree nut allergy are not uncommon. Allergy and intolerance to other foods often improve with time and avoidance. It may be reasonable to suggest that an open challenge be undertaken at six-monthly intervals with a small amount of the suspected food. This can be done at home unless there is a history of severe food allergic reaction. In children with a history of severe reaction to egg or cow's milk, the challenge should be performed in the hospital or where facilities for resuscitation are available.

5.2.2 Approach to food avoidance

Once the diagnosis of food allergy or intolerance has been made, avoidance of the offending food (or foods) is the most important treatment.[5] The diagnosis of food intolerance is not always easy, as patients often tend to blame foods for

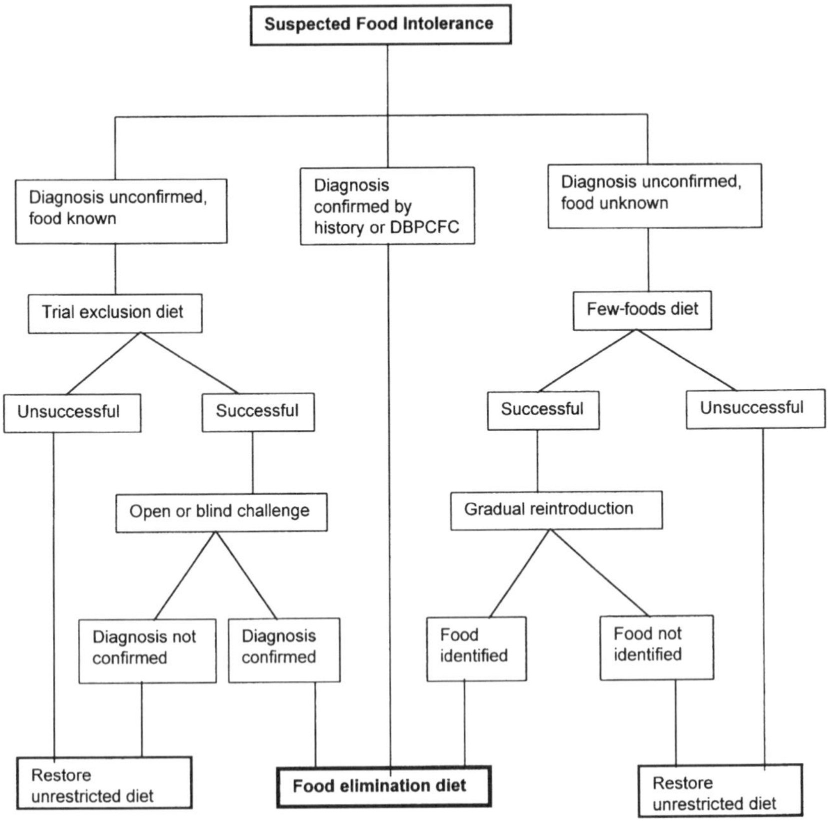

Fig. 5.1 Algorithm for the management of food tolerance.

many symptoms that are both related and unrelated to food and there is no test which can reliably exclude or confirm the condition. The approach to management of food intolerance depends on the clinical situation (Fig. 5.1).

Confirmed food intolerance

If the history is highly suggestive of food allergy or intolerance, for example where an acute reaction has occurred immediately after ingesting a food, further confirmation may not be required. The suspected food should be excluded from the diet. If the subject is highly allergic, small amounts can produce severe, and even fatal, allergic reactions.[3] A dietitian is best placed to give explanation and suggest alternatives. Prophylactic pharmacotherapy, if needed, should also be prescribed. Highly atopic children and occasionally adults may be allergic to several foods such as cow's milk, egg and nuts. It is important that the diagnosis is made accurately and careful consideration is given to providing information about alternative foods with adequate replacement of essential nutrients.

In adults tolerance is less likely and foods should be avoided for life. In children, tolerance often develops to foods such as milk and egg and occasionally nuts, and an open challenge may be warranted after a period of exclusion. This should be done in a hospital setting if the reaction was categorised anaphylactic. The results of immunological tests (skin test or RAST), and any reaction to inadvertent exposure, guide the need for challenge.

Unconfirmed food intolerance (food known or suspected)
If the food is known or suspected either from the history or from skin test or RAST, a trial exclusion diet is recommended. The period of exclusion depends, to some extent, on the type and frequency of reaction. If the subject is having frequent symptoms, for example urticarial episodes several times a week or diarrhoea, exclusion for a couple of weeks might be sufficient to gauge the response. For atopic eczema or chronic recurrent urticaria, a longer period, i.e. 2–3 months, may be required to assess improvement, allowing for spontaneous fluctuation in the disease severity.

Monitoring of symptoms should be as objective as possible. The patient should keep a record of daily or weekly symptoms in a diary. If no improvement occurs, unrestricted diet should be restored and the patient reassured that the condition is not related to foods.

If significant improvement is observed on a trial exclusion diet, then open or DBPCFC should be performed.[6] An open challenge or reintroduction of the food in the diet may be sufficient if an objective improvement has been observed, for example if the frequency and severity of urticaria or severity of eczema assessed objectively with a standardised score has reduced. If reintroduction causes a relapse of the disease, diagnosis is confirmed and the food should be excluded for a longer period. Tolerance to food often occurs after a period of avoidance, especially in children, and it is important to reintroduce or challenge at 6–12-monthly intervals.

If the symptoms are largely subjective, such as headaches or behaviour changes, and an improvement is observed with the exclusion diet, DBPCFC is essential. If the diagnosis is confirmed, food should be excluded for a longer period. Again, reintroduction or challenge should be done at regular intervals, especially in children who often grow to tolerate the food. When more than one food is suspected the dietitian should carefully monitor the exclusion diet and suggest appropriate replacement foods.

Reintroduction or challenge following improvement is more complicated and best done sequentially, selecting the food most likely to cause the patient's symptoms. A full medical and dietary history is of value and skin tests or RAST may also be helpful.

Unconfirmed food intolerance (food unknown)
Sometimes the history suggests, or the patient is convinced, that the symptoms are due to food intolerance but the food is not known. In these situations, a few-foods diet is helpful. A set of a few foods is selected which should include at least one source of carbohydrate and proteins. Supplements may be required for minerals, trace elements and vitamins. If there is no improvement, another set of

a few foods may be tried with a different source of carbohydrate and proteins. If there is still no improvement, the patient should be reassured that foods are not involved.

If significant improvement occurs, a group of foods such as cereals or meat may be introduced at intervals. The period of assessment depends on the nature and frequency of symptoms. If symptoms return on introduction of a food or a group of foods, then this is excluded and the process continues. In this way one or more foods are identified and open or double-blind challenge should then be performed to confirm the diagnosis. This process is tedious and should only be undertaken when symptoms are severe and interfere with the patient's life. The help of a qualified dietitian is mandatory.

5.2.3 The role of the dietitian

Dietary history

A careful dietary history should reveal the patient's dietary habits, their intake of the suspected food and the extent to which they have to modify their diet to comply with the dietary avoidance. It can also help to identify a situation where accidental exposure is likely to occur. Likes, dislikes and cravings for foods are noted. There may be preconceived ideas about various foods and some foods may already be excluded at the patient's own initiative or following advice from other medical or non-medical practitioners. A food diary completed over a week could be analysed by a computer program to give an indication of the patient's dietary intake. An assessment is also made of the patient's ability to understand and comply with a possibly difficult avoidance diet. This information would help in the diagnosis and assessment of risk to the patient and is also valuable when an avoidance diet is suggested.

Avoidance diet

The dietitian should give an explanation and provide a list of packaged foods that might contain the food to be excluded. Replacements should be suggested, such as an alternative cereal or meat. For example, in patients with wheat or gluten intolerance, the dietician would advise on the availability of wheat- or gluten-free bread and other products. A list of food free from the food allergen is very helpful. Patients should be told how to read labels (such as 'E' numbers) and what to look for when buying packaged foods or eating in a restaurant.

Nutritional content

It is the responsibility of the dietitian to make sure that the avoidance diet is nutritionally adequate. This should be done with respect to intake of energy (calorie value), proteins, fibre, calcium, vitamin, iron and other minerals. Supplements should be suggested if alternative sources are liable to be inadequate in a particular nutrient. For example, milk is an important source of calcium. Therefore, in a milk or dairy food avoidance diet, supplements of calcium should be prescribed.

Follow-up

An important role of the dietitian is to provide support during the trial exclusion diet and during the early periods of the avoidance diet when the patient may not be confident of their ability to successfully avoid the food. Telephone, postal or e-mail support for answering any questions between clinic follow-ups is highly appreciated by the patients and their relatives.

5.3 Hypoallergenic foods

5.3.1 The need for hypoallergenic foods

Food proteins are essentially foreign proteins capable of eliciting immunological responses. Any food protein may be allergenic if it can be absorbed intact or as substantial fragments, through the gut mucosa, and then evoke an immune (allergic) response. Some foods, such as rice and vegetables, are less allergenic than others, such as milk, egg and nuts. The intrinsic properties of the protein, the overall composition of the food, and the processing (especially thermal processing) all have an effect on the allergic potential. In the management of food allergy it is possible to exclude the food responsible for symptoms and to replace it with less allergenic foods. In certain situations it is not possible simply to eliminate the food, e.g. milk during infancy. Up to 2.5% of infants are affected by cow's milk allergy (CMA) in the first two years of life, although most of these children will outgrow their reactivity within 2–3 years. However, during the interim period an alternative milk formula is usually required.

5.3.2 Types of foods available

Allergic reactions require large protein molecules (antigens) to stimulate the production of antibodies. To reduce allergenicity, the source protein can be broken down into small peptide molecules and amino acids by enzyme hydrolysis. This process has been used successfully in the production of hydrolysed formulae (HF). These infant formulae are based on animal or vegetable protein (casein, whey, soy and bovine collagen) and are used extensively in children with cow's milk allergy or intolerance.

In gluten-induced enteropathy a specific protein (gluten) is responsible for stimulating the immune reaction. Foods have been prepared without gluten, that are suitable for these individuals. When a protein is denatured by heat, most of the original tertiary structure is lost, so that many of the sites recognised by antibodies on the native molecule are destroyed. There are many examples of allergenicity being reduced, but not eliminated, by heating. Thermal processing can be part of a procedure for making hypoallergenic food, but will rarely be sufficient on its own.

5.3.3 Hypoallergenic milk formulae[7]

Indications
CMA in the first year of life is one of the most common problems faced by paediatricians. It is mediated by an immune mechanism, whereas cow's milk intolerance is due to non-immunological causes such as lactase deficiency. CMA may affect the gastrointestinal tract, respiratory tract, skin or blood, and systemic reactions, including anaphylaxis, may occur. Avoidance is the mainstay of treatment, and breast-feeding should be actively encouraged. Since intact cow's milk protein can pass into the breast milk, the lactating mother should avoid the excessive intake of milk products herself and take a calcium supplement. If breast-feeding is not feasible or if supplements are required, soya milk, hydrolysate or amino acid-based formulae may be used.

Hydrolysed formulae
According to the definition of the European Scientific Committee for Food, hypoallergenic or hypoantigenic formulae are those which contain hydrolysed protein. The peptides of HF should be as short as possible. In extensively hydrolysed formulae (eHF) 95% of peptides have a molecular weight below 1500 dalton and less than 0.5% of the remaining peptides are above 6000 dalton. Partially hydrolysed formulae (pHF) have 2–18% of peptides above 6000 dalton. These larger peptides may elicit allergic reactions. pHF have a higher capacity to induce positive skin tests and provocation tests and to bind to the human serum IgE antibodies of children allergic to cow's milk. Amino acid-based formula does not have peptides so there is no likelihood of allergic reactions.

ELISA inhibition assay, with polyclonal antibodies specific for casein components of cow's milk, is a sensitive method for estimating residual antigenicity in hypoallergenic infant formulae, suggesting their potential application for quality control. Some HF are not optimal in their nutritional content. The process to reduce allergenicity may modify amino acid content or reduce its bioavailability. Changes in the absorption of calcium, zinc and copper have been found. All infant formulae promoted as 'hypoallergenic' should also be tested in milk-allergic patients to assess their allergenic potential, in addition to standard nutritional evaluation and laboratory and animal testing for antigenicity.[8]

Choice of formula
The choice of the substitute milk depends on its allergenicity, nutritional composition, palatability and cost. Soya milk may be safely used in many children with CMA. However, 5–30% of children with CMA are also allergic to soya protein, and some children with CMA become allergic to soya milk after its introduction. eHF have been used extensively for the treatment of children with CMA and are generally well tolerated, although there are several reports of allergic reactions, including anaphylaxis. pHF are more palatable but, because of their higher allergenicity, they are not generally recommended for the treatment of CMA.

It is recommended that children with CMA should be skin tested with eHF before this is prescribed. A negative reaction indicates eHF is safe to use. Children with a positive skin test result to the eHF should be further evaluated by an open challenge in a hospital setting where facilities for resuscitation are available.

5.3.4 Prevention of allergy

The need for prevention

There is general consensus that the prevalence of asthma and other atopic diseases, including food allergies, is increasing. A history of allergic disease in the immediate family (atopic heredity) is the most important risk factor. Recent studies indicate that exposure to allergens *in utero* and in the first few months of life is critical in the development of allergic disease in children with an atopic heredity. In children at high risk, reduction in exposure to allergen should lead to a decline in disease prevalence. Food proteins are important allergens in early childhood. A hypoallergenic diet has therefore been suggested as a means of preventing the development of allergy.

Use of hypoallergenic diet in prevention

Experimental evidence indicates that the child can be sensitised *in utero*. It is sometimes advised that an atopic mother should avoid highly allergenic foods during pregnancy. However, there is concern that this might adversely affect the growth of the foetus. Avoidance of allergens during early infancy has been shown to reduce the development of allergy in at-risk infants. Among food allergens, cow's milk is an important allergen at this stage, and exclusive breast-feeding has been advocated. As protein ingested by the lactating mother can be secreted in the breast milk (a potential source of sensitisation), a maternal diet excluding allergenic foods during lactation has been advised. eHF may be used, if required, as a replacement or supplement to breast milk and by pregnant and lactating women if cow's milk is excluded from their diet. There are, however, problems designing suitable hydrolysates that are low in antigenicity and palatable in taste.

Six months' delay in the introduction of solid foods and a further 1–2 years' delay for more allergenic foods such as eggs, fish and nuts have also been recommended. This requires supervision by a qualified dietitian so that replacement foods are suggested and nutritional adequacy of the diet is ensured. Any primary preventive programme for infants at high risk requires highly motivated parents and close cooperation with the physician and other healthcare workers.

Outcome

In infants at risk of developing allergies, maternal avoidance of hypoallergenic foods during lactation, exclusive breast-feeding for 4–6 months, use of eHF if required, and introduction of solids after four months of age reduce the incidence of atopic syndromes, particularly atopic dermatitis and food allergy in early childhood. There is also evidence of reduced sensitisation, i.e. the number of positive skin prick tests and level of specific IgE antibody.[9]

5.4 Drug treatment[10]

The treatment of choice in intolerance to food is avoidance. Pharmacological agents play a secondary role in management. Pharmacotherapy is indicated:

- for unintentional exposure to the food,
- when symptoms persist despite efforts to maintain an allergen-free diet, and/or
- where identification of the responsible food is not possible.

5.4.1 Adrenaline

Subcutaneous or intramuscular adrenaline is used as the first-line treatment for anaphylactic reaction to food and other allergens.[11] The intramuscular route is preferable if there is evidence of circulatory collapse, as the absorption is better than from the subcutaneous site. Patients who are at risk of anaphylactic reactions, for example those with nut allergies, should be provided with a self-injectable adrenaline device. This delivers a set dose of adrenaline by intramuscular route. The adult dose is 300 μg and the paediatric dose is 150 μg; repeatable after 15 minutes. Patients and their carers should be given instructions in the use of the device in case of emergency. When absorption from the intramuscular route is not adequate, for example in severe hypotension and shock, slow intravenous injection may be used by trained personnel. Inhaled adrenaline is not useful for the treatment of anaphylaxis. However, it may be effective for angioedema or laryngeal oedema in the absence of systemic symptoms.

5.4.2 Antihistamines

Introduction

Antihistamines interfere with the binding of histamine to its receptors. There are three types of antihistamine receptors: H_1, H_2 and H_3. H_1 receptors are important in allergic reactions and their blockade by antihistamines reduces symptoms such as itching, rash and vasodilation. These are absorbed rapidly from the gastrointestinal tract and metabolised in the liver.

H_1 receptor antagonists

The classical antihistamines such as chlorpheniramine are effective H_1 blockers but sedation is prominent. There is little evidence to suggest that one antihistamine is better in effectiveness than others, though individual response may vary widely. The duration of action and side-effect profile may determine

Table 5.4 Commonly used sedative and non-sedative antihistamines

Sedative antihistamines	Non-sedative antihistamines
Chlorpheniramine	Loratadine
Clemastine	Cetirizine
Hydroxyzine	Fexofenadine
Promethazine	Acrivastine
Cyproheptadine	Terfenadine
Azatadine	Astemizole
Brompheniramine	
Trimeprazine	

the choice. The second-generation antihistamines are at least equally effective and are much less sedating (Table 5.4).

Indications

In the treatment of food allergy, antihistamines are given primarily to relieve symptoms such as itching and urticaria due to inadvertent exposure. Oral symptoms, such as itching in the mouth and throat and swelling, may also respond but there is little effect on gastrointestinal symptoms such as vomiting and diarrhoea. For mild symptoms, oral antihistamine may be effective and may be continued until symptoms disappear. For moderate to severe allergic reactions, antihistamine should be given through the parentral route for rapid systemic availability. Occasionally antihistamines are used regularly for chronic food allergic symptoms where causative food(s) have not been identified.

Side effects

Drowsiness and antimuscarinic effects such as urinary retention, dry mouth and blurred vision are major disadvantages with older antihistamines. The so-called non-sedating antihistamines can also cause drowsiness in some patients. Arrhythmias may occur in high doses, particularly with terfenadine and astemizole.

5.4.3 Cromoglycate

Sodium cromoglycate is a sodium salt of chromone-2-carboxylic acid. It inhibits the release of mediators from mast cells and basophils, although this does not fully explain its effectiveness in IgE-mediated allergic diseases. Only 1% of the orally administered dose is absorbed from the gastrointestinal tract.

Oral sodium cromoglycate may be useful in some patients with multiple food allergies.[12] It is a less effective but safer alternative to steroids in the management of chronic food allergy not responding adequately to food allergen avoidance. However, it should not be used in place of allergen avoidance. Acute symptoms such as bronchospasm, rash, nausea and diarrhoea respond better than do chronic food-related diseases such as atopic eczema. Side effects are minimal, although nausea, rashes and joint pain have been reported.

Table 5.5 Adverse effects associated with glucocorticoid and mineralocorticoid activity of corticosteroids

Activity	Side effects
Mineralocorticoid	Hypertension, sodium and water retention, potassium loss and adrenal suppression
Glucocorticoid	Diabetes, osteoporosis, psychosis, proximal myopathy, peptic ulceration, cataract and skin atrophy, hirsutism, reduction in ability to fight infection

5.4.4 Corticosteroids

Corticosteroids are indicated in severe systemic allergic reactions when intravenous hydrocortisone may be needed followed by a short course of oral steroids. Rarely, in patients with severe atopic allergy to multiple foods or where causative food(s) are not known and symptoms are severe, systemic steroids can be used on a long-term basis. The dose should be kept to a minimum because of the well-known side effects. These side effects can be explained by the mineralocorticoid or glucocorticoid activity (Table 5.5) of these substances. Different steroids vary in their glucocorticoid (anti-inflammatory) and mineralocorticoid (water retaining) effects. Hydrocortisone is less potent than others in its anti-inflammatory activity but is administered by intravenous injection for acute systemic reactions in addition to adrenaline. Prednisolone has more potent glucocorticoid activity and is often used if oral therapy is required. Other steroids such as betamethasone and dexamethasone have very high glucocorticoid activity and may be used for long-term therapy.

5.4.5 Other symptomatic treatment

Ketotifen with antihistaminic and anti-inflammatory properties has been used in food allergic reactions such as urticaria and bronchospasm. It may be useful as an additional therapy in some patients. Beta-2 agonists such as salbutamol or terbutalin may be used when bronchospasm is a prominent feature in an allergic reaction. These drugs can be delivered by inhalation through a metered dose inhaler, in an aerosol form through a nebuliser, or by intravenous route. Food-related eczema and rhinitis should be treated along the standard line with topical steroids and antihistamine in addition to allergen avoidance.

5.4.6 Specific immunotherapy

Immunotherapy (desensitisation) has been used in the treatment of allergic diseases since 1911. Extracts of allergen to which the patient is sensitised are given in increasing concentration, starting with a very dilute solution, until tolerance is achieved. Allergen immunotherapy is specific to the allergen being administered. The exact mechanism is unknown but presumably depends on the development of specific IgG antibodies which bind to the IgE receptor on mast cells and basophils, thus making it unavailable for IgE. This mode of treatment has been used successfully for the treatment of pollen and insect allergy but its usefulness in other allergic diseases has been controversial.

Several studies evaluating the effectiveness of specific immunotherapy in food allergic diseases such as peanut and fish allergy produced conflicting results. The majority of the studies did not find evidence of protection in peanut allergic patients, and severe reactions during the treatment were common. However, some studies have supported the use of immunotherapy in the treatment of fish and egg allergy. The overall consensus is that specific immunotherapy has no place in the treatment of food allergy.

5.4.7 Novel treatments

IgE-mediated food allergic reactions depend on the binding of allergen with IgE antibodies bound to the receptors on the mast cells and basophils. This antigen–antibody reaction causes degranulation of mast cells with the release of preformed and newly synthesised mediators initiating a cascade of inflammatory cell influx, and production of cytokines and mediators. The clinical effect of this is an allergic reaction that may be systemic (anaphylactic reaction), or localised to an organ or tissue (allergic inflammation/disease).

Peptides have been synthesised which are able to bind to the IgE receptors. This will competitively inhibit IgE binding to the receptor. These peptides, therefore, have the capacity to block IgE-mediated reactions non-specifically. They may be useful in patients with multiple food allergies or other IgE-mediated diseases such as asthma and rhinitis. Several such peptides are in the developmental phase.

A novel method of blocking the IgE-mediated reactions is to develop monoclonal antibodies to the IgE molecule. These anti-IgE antibodies bind to the free IgE in the circulation, thus reducing the available IgE to bind to mast cell receptors. In clinical studies, anti-IgE antibodies have been found to be useful in allergic asthma and several large-scale studies are being conducted. It remains to be seen if anti-IgE antibodies will be useful in food allergic disorders such as peanut, egg and cow's milk allergy.

5.5 Treating the immediate symptoms

5.5.1 Acute allergic reactions to foods

Development of symptoms within two hours of ingestion of the suspected food may be reasonably classified as an acute reaction. These reactions are commonly due to milk, egg, fish and nuts (Table 5.6). The person may or may not know the food responsible. In children, allergic reaction may occur to the first known exposure to a food such as cow's milk, egg or peanut. It may also develop in an

Table 5.6 Foods commonly implicated in the anaphylactic reactions

Peanut
Tree nuts: brazil nut, hazelnut, almond, walnut, pistachio, cashew nut, pecan
Cow's milk
Soya milk
Fish/shellfish
Egg

adult to a food previously well tolerated although this is uncommon. Acute allergic reactions are usually IgE mediated.

Mechanism
Allergic reactions occur as a result of interaction of allergen with IgE antibodies bound to receptors on the surface of mast cells. This interaction results in the release of mediators such as histamine, heparin, bradykinin, prostaglandin and leukotrienes. The allergen may come from a variety of sources such as foods (e.g. peanut), drugs (e.g. penicillin), insects (e.g. bee venom), etc. The reaction may involve one or more systems and may be mild, moderate or severe.

Assessment of severity (mild, moderate and severe)
The severity of reaction depends on sensitivity of the patients to food allergen and the amount ingested. Patients with life-threatening symptoms such as respiratory difficulty due to laryngeal oedema or severe bronchospasm and/or hypotension should be regarded as having a severe reaction or anaphylaxis.[13] Troublesome, but not immediately life-threatening, reactions such as generalised urticaria/angioedema and bronchospasm of moderate severity may be termed severe allergic reactions. Sometimes the reaction is mild and confined to an organ or system, for instance oral or gastrointestinal symptoms or localised urticaria.

5.5.2 Treatment of anaphylaxis

Foods are the commonest cause of anaphylaxis and there appears to be a rise in its prevalence.[14] There is no universally agreed definition of anaphylaxis, as symptoms of acute allergic reaction may vary widely. Allergic reactions with one or more life-threatening features may be regarded as anaphylaxis. There have been several reports of deaths due to anaphylactic reactions to foods, especially peanuts and tree nuts. A history of previous anaphylactic reaction to a food is the most important risk factor for the prediction of future anaphylaxis.

Clinical features
A wide range of symptoms has been observed in allergic reactions to foods (Table 5.7). The type and severity of symptoms depend on the patient's sensitivity to the food, the amount ingested and the route of entry of the

Table 5.7 Clinical features of anaphylactic reaction; not all symptoms are present in every patient

Erythema, rash, generalised itching
Numbness and tingling of lips and mouth, swelling of tongue
Urticaria, angioedema
Bronchospasm, wheezing, sneezing
Throat tightness, stridor (laryngeal oedema)
Nausea, vomiting, abdominal pain
Tachycardia, palpitation, arrhythmia
Feeling unwell with a sense of doom, dizziness, fainting
Hypotension, collapse, shock, loss of consciousness, death

allergenic food.[14] Those who are highly sensitive may react to even smell and/or touch of the food. Local reactions such as urticarial rash and itching may occur even on touching the food. Children who are highly allergic to milk, egg and nuts may react to contact with these foods, with erythema, rash and localised swelling. Oral symptoms, including numbness and itching on the lips, tongue and inside of the mouth, occur within minutes of contact with the oral mucosa. There may also be generalised pruritis, urticaria, swelling of the lips and tongue, feeling unwell and a sense of doom. This may progress to laryngeal oedema (causing upper airway obstruction and respiratory difficulty), hypotension, loss of consciousness and shock. The onset of action may be rapid, and a previously well person can become moribund within minutes of ingestion of food.

Treatment

Adrenaline given subcutaneously or intramuscularly is the first-line therapy. The treatment with adrenaline is not without risk but should not be held back if it is needed. It is the most useful drug to counter the dangerous effects of large amounts of histamine and other mediators released into the bloodstream.[15] Absorption is better through an intramuscular route. The usual dose is 0.3–0.5 ml of 1:1000 solution. The dose can be repeated after 15–30 minutes if response is not adequate. The condition invariably responds to adrenaline if given early during the reaction. Therefore, early treatment is crucial and may often need to be given outside the hospital setting. However, treatment with adrenaline should not be relied upon as the complete treatment, and medical help should always be sought.

In a hospital setting intravenous adrenaline may be used, as a dilute solution (1:10 000) given slowly with cardiac monitoring, in those who do not respond to adrenaline by intramuscular route. However, it may be difficult to find venous access in a collapsed patient and side effects are more likely to occur. Other therapies include antihistamine such as chlorpheniramine (10 mg given intravenously), followed by intravenous steroids such as hydrocortisone (100 mg). An assessment should be made of the circulatory status, and intravenous fluids (colloids) are administered. Treatment of bronchospasm (nebulised bronchodilators) and arrhythmia (anti-arrhythmics) may be required. Oxygen may be required for patients with respiratory symptoms.

Long-term management

The patient should be referred to an allergy specialist to establish the cause of anaphylaxis and appropriate advice on management.[13] A detailed history is taken and a skin prick test performed to confirm the IgE-mediated allergy. The cause is sometimes obvious, as in patients who react immediately after eating a nut. Others may have a history of anaphylactic episodes after a meal and the food responsible for the reaction may be more difficult to identify.

It is important that patients are prepared for accidental exposure. Treatment with adrenaline at the onset of a reaction is effective and the patient should be given a syringe loaded with adrenaline for self-injection or injection by friends or relatives. This is available in two strengths: for adults, 0.3 ml of 1:1000 solution (0.3 mg) and for children, 0.3 ml of 1:2000 solution (0.15 mg). The dose can be repeated after 15 minutes if the response is inadequate. Patients and their carers should be instructed in the use of these syringes. For children it is important that schools are aware of the problem and know when and how to use the adrenaline treatment. It has been commonly observed that patients or children's parents are provided with a pre-loaded syringe without adequate explanation and instructions in its use. This has resulted in heightened anxiety, which may be worse than the fear of anaphylaxis itself. It is also important to stress that the availability of adrenaline is not a substitute for food avoidance.

5.5.3 Severe allergic reactions

Following assessment of severity, if the reaction is not thought to be immediately life-threatening, antihistamine and hydrocortisone may be given while the patient is observed in a medical facility.

Generalised urticaria/angioedema

This can be quite dramatic, with erythema and rash all over the body surface, and swelling of the face, lips and tongue. However, if confined to the skin and oral mucosa, it is usually not life-threatening. This reaction responds to oral or parenteral antihistamine in addition to corticosteroids. Treatment may need to be continued for a few days until symptoms have completely subsided. Unless the cause of the reaction is known, the patient should be referred to an allergy clinic for evaluation.

Bronchospasm

Mild to moderate bronchospasm commonly occurs as part of a generalised reaction but may be the most prominent symptom. This usually responds to inhaled or nebulised bronchodilator in addition to corticosteroids. Short-acting beta-2 agonists such as salbutamol or terbutalin may be given through a metered dose inhaler attached to a spacer device when outside the hospital. If the response is not adequate, the patient should be transferred to the hospital where nebulised or intravenous bronchodilator may be administered. Corticosteroids are also used, but their onset of action is delayed for a few hours and is not immediately beneficial. Oxygen should also be given as hypoxia may occur.

5.5.4 Treatment of mild or localised reactions

A mild form of urticarial rash and itching may be the only manifestation if the sensitivity to food is low or only a small amount has been ingested. Oral symptoms, of swelling and numbness of the lips and localised itching, are common symptoms of allergy to fresh or raw fruits in some patients who are highly sensitive to pollen (oral allergy syndrome). Treatment with oral antihistamine may be sufficient for these episodes. Patients should keep a supply of non-sedating antihistamine such as cetirizine (10 mg) or loratidine (10 mg) tablets. For children, antihistamine syrup (cetirizine or chlorpheniramine) should be prescribed. If the episode does not respond to oral antihistamine or if there are signs of progression, medical help should be sought.

5.6 Treatment of common food allergic diseases

5.6.1 Gastrointestinal symptoms

Diarrhoea, vomiting and abdominal colic are common manifestations of food intolerance but may also be due to infective or other causes. The cause should be established by appropriate investigations. Food intolerance causing gastro-intestinal symptoms could be due to enzyme deficiency and immunological and non-immunological reactions to foods. Cow's milk intolerance is a common problem during infancy that can be treated by excluding cow's milk from the diet. Replacement with soya milk or hydrolysed formula is given. Secondary lactose deficiency is relatively common following gastroenteritis, which is self-limiting. Avoidance of milk and milk products is essential during this period. In adults, some cases of irritable bowel syndrome may be due to food intolerance. If one or more foods is suspected this can be excluded from the diet and the response observed.

5.6.2 Gluten-sensitive enteropathy

Gluten-sensitive enteropathy or coeliac disease is a malabsorption syndrome due to lymphocyte-mediated hypersensitivity to storage proteins found in wheat and some other cereals. The most important of these is gliadin. Exclusion of gluten-containing foods (wheat, barley, rye and oats) from the diet leads to improvement in symptoms. Tolerance does not develop and avoidance is lifelong. Gluten-free products should be prescribed for patients suffering from this disease (Table 5.8).

Table 5.8 Some wheat- and gluten-free products available on prescription and over the counter

Product name	Products
Bread/rolls/pizza bases	
Glutafin	Multigrain range loaf
	Pizza bases
Schar	Gluten-free bread and rolls
Flour mixes	
Glutafin	Gluten-free fibre or white mix
Schar	Bread and flour mix
Glutano	Flour mix
Juvela	Gluten-free harvest mix
Trufree	Flour mix
Biscuits/crackers	
Glutafin	Crackers, savoury biscuits, sweet, digestive and tea biscuits
Schar	Crispbread, crackertoast, sweet biscuits
Glutano	Crackers, gluten-free biscuits
Pastas	
Glutafin	Wheat and gluten-free pasta
Schar	Wheat and gluten-free pasta
Glutano	Wheat and gluten-free pasta

5.6.3 Atopic eczema

A proportion of young children with atopic eczema show an improvement when selected foods are excluded from their diet.[16] Common foods implicated in the causation of eczema are egg and milk, and in some cases wheat and peanut. A detailed history and skin prick tests or RAST are sometimes helpful in identifying the food, but a negative test does not exclude the possibility of benefit from a food exclusion diet. If the child is sensitised to one or more foods on skin test or RAST, a trial diet excluding these foods should be prescribed for 4–6 weeks. If the child is not sensitised to foods, a trial of cow's milk and egg may be of value in children with extensive eczema.

Eczema is a chronic disease and improvement with exclusion diet may not occur immediately. An open challenge should always be undertaken if an improvement has been observed to confirm the causative relationship. Double-blind challenge may not always be feasible in clinical practice. Once the food(s) are identified, a longer period of dietary avoidance is undertaken. Cow's milk could be replaced with soya milk (if the child is not allergic to soya) or hydrolysed formulae. Supervision of a qualified dietitian is recommended to ensure compliance and nutritional adequacy of the diet. As food allergy and eczema improves during early childhood, an open challenge may be undertaken every 6–12 months depending on the severity of eczema and results of the skin test or RAST.

The benefit of the diet should be balanced against the risks and inconvenience of a restricted diet. Therefore, dietary treatment may not be suitable for mild eczema patients who will often respond to the regular use of emollients. Also the diet should always be used in conjunction with other standard treatment. If no improvement is observed on the trial diet after a period of 6–8 weeks, the diet should be discontinued. In a small number of children with very severe eczema, a few-foods (elemental) diet could be tried for a short period. If a significant response occurs, the foods should be sequentially reintroduced in an attempt to identify the food(s) responsible for causing eczema. In young children, a few-foods diet should be very carefully monitored by a qualified dietitian. Adult-onset eczema is rarely food-related and usually does not respond to an exclusion diet.

5.6.4 Urticaria and angioedema

In some patients, chronic recurrent urticaria and angioedema can be due to intolerance to food additives.[17] Other causes include drugs, physical factors (cold, pressure, heat, etc.) and foods. However, it should be remembered that no cause can be found in the majority of patients (idiopathic chronic recurrent urticaria). An attempt should be made to identify a cause. If a drug or a food is found to be causative, this should be excluded. In cases where no cause is obvious, the best approach is to treat the condition with the regular use of antihistamine for a period of 6–12 weeks. If the response to antihistamine is not adequate or if symptoms return once antihistamines are discontinued, a trial diet excluding benzoate preservatives, azodyes and salicylate may be helpful. If the condition improves, an open or double-blind challenge may be warranted to confirm the causative relationship. If positive, the diet should be continued for a longer period until spontaneous remission occurs. If the condition does not improve or the challenge does not confirm an association, the diet should be discontinued.

5.6.5 Diet and behaviour problems

In children, food intolerance has been implicated in hyperactivity or behaviour disorders. It has been claimed by some that children's behaviour improved on a food diet containing few or no additives. Parents often strongly believe that food additives, especially azodyes (e.g. tartrazine), are responsible for the child's behaviour. However, scientific proof is lacking and studies have not been able to prove conclusively the effectiveness of the dietary approach. There are considerable problems in subjecting children to a restrictive diet where behaviour problems already exist. At the current state of knowledge, a dietary approach is not recommended for this common problem.

5.6.6 Respiratory symptoms

Food is an uncommon trigger for upper or lower respiratory symptoms. IgE-mediated allergy to foods such as nuts, fish, egg, etc., may cause respiratory difficulty and bronchospasm as part of a systemic reaction. Mild laryngeal symptoms are more frequently seen in food allergic patients and include throat tightness, itching in the throat or a dry cough. Patients at risk of systemic reactions should be very vigilant in avoidance and may need to carry a pre-loaded adrenaline syringe if there is a history of severe bronchospasm or respiratory difficulty. Some highly atopic patients with atopic dermatitis, in addition to asthma or rhinitis, may develop respiratory symptoms following ingestion of foods. This should be established, by history or double-blind challenge if required, and appropriate avoidance should be practised.

Sulphites are usually added to foods (vegetables and fruits) and soft drinks as preservatives or found naturally in beer and wines.[18] Ingestion or inhalation of sulphites may cause bronchoconstriction in some asthmatic individuals. A careful history should reveal the cause and appropriate advice regarding avoidance should be given.

5.6.7 Migraine and headaches

Occasionally foods with a high content of tyramine, such as cheese, coffee, red wine and yeast extract, are responsible for migrainous headaches.[19] In some patients the association is obvious and these patients usually avoid these foods. In other cases of chronic headache, once other treatable causes have been excluded, a diet excluding foods with high tyramine content may be tried. However, double-blind challenges are often unsuccessful in confirming a relationship of foods with headaches.

5.6.8 Allergy to food additives[20]

An additive is a substance added to foods for preservation, coloration and some other purposes. Additives are numerous and include benzoates, metabisulphites and azodyes. The prevalence of adverse reaction to additives is 0.03–0.5%. Adverse reactions to additives occur in 20–25% of patients with aspirin intolerance and in 10–20% with chronic recurrent urticaria. IgE-mediated hypersensitivity, resulting in acute allergic reaction, has been described for azodyes, ethylene oxide and penicillin, and delayed-type hypersensitivity for nickel salt. A list should be provided of foods containing the additive that the patient does not tolerate. Clear labelling of packaged food helps to avoid accidental exposure.

5.7 Summary: trends in treatment

Despite recent advances in our knowledge of immune processes involved in food allergy and intolerance, there have been few major developments in the treatment of this common condition. Avoidance of the offending food remains the mainstay of treatment. Pharmacological therapy is useful in acute reaction due to inadvertent exposure but generally disappointing in the treatment of chronic food intolerance.

The importance of a detailed history cannot be overstated. The diagnosis can often be made on the history alone. A dietary history helps to identify the consumption of the offending food and aids in suggesting replacement. Patients with a history of acute allergic reaction to foods such as milk, egg, fish or nuts have to be extremely careful in consuming packaged food or when eating out. Packaged foods should be labelled clearly with the highly allergenic foods to reduce avoidable morbidity and mortality.

In children with cow's milk intolerance, the development of relatively safer extensively hydrolysed formulae has been a welcome relief. However, the increasingly complicated processing of foods may unravel new food antigens. Indeed, the increase in the prevalence of allergic diseases in general and of food allergy in particular had been blamed on the drugs and fertilisers used by farmers and the addition of an ever-expanding list of preservatives used by the food industry. More recently there has been considerable debate on the usefulness or otherwise of genetically manipulated foods.

It is not always possible to make a confident diagnosis of food intolerance on history alone. A trial exclusion of the suspected food may lead to an improvement in symptoms. Careful monitoring of symptoms during the trial diet, preferably with the help of a symptom diary, is essential. A dietitian's services are invaluable in organising a trial exclusion diet. Where food intolerance is suspected and symptoms are severe but the food is not known, a trial of a few-foods or elemental diet may be warranted. The diagnosis should be confirmed by DBPCFC, where possible, before a longer period of avoidance is recommended, as placebo responses are not uncommon. The dietitian can also provide written and verbal explanation of the avoidance measures and ensure that the recommended diet is nutritionally adequate. Assessment by the dietitian may reveal a need for supplements of calcium, vitamins or a different source of protein or calories.

In children, allergy to multiple foods is common, and appropriate avoidance, in addition to replacement where necessary, leads to improvement in symptoms. This is usually self-limiting and children tend to grow out of the allergies. In adults, intolerance to multiple foods is rare. It is important that the avoidance diet contains alternative sources of protein and calories and appropriate supplements are provided. Prophylactic treatment with drugs such as antihistamine and cromoglycate is occasionally useful. Manipulation of the immune system to alter its response to food allergen specifically to a food (specific immunotherapy) has not been very successful. Non-specific immunotherapy with peptide or DNA vaccines is being studied. An alternative approach is to reduce the antigenic component of the food to manufacture hypoallergenic foods.

Acute allergic reactions with life-threatening features are treated effectively with adrenaline. Patients at risk of these reactions should carry pre-loaded adrenaline at all times. Milder forms of acute reactions may respond to antihistamines. Prophylactic treatment of food allergy, to prevent an acute reaction, is not satisfactory. Recent development of drugs such as monoclonal antibodies to IgE, which inhibits all IgE-mediated allergic reactions, may prove to be useful in the prophylaxis of acute reactions or in the treatment of chronic food allergic symptoms.

5.8 Sources of further information and advice

Anaphylaxis Campaign
PO Box 149, Fleet, Hampshire GU13 0FA
Tel. 01252 542029

British Allergy Foundation
Deepdene House, 30 Bellegrove Road, Welling, Kent DA16 3PY
Tel. 020 8303 8525

Latex Allergy Support Group
PO Box 36, Cheltenham GL52 4WY

National Asthma Campaign
Providence House, Providence Place, London N1 0NT
Helpline: 0845 701 0203

National Eczema Society
163 Eversholt Street, London NW1 1BU
Tel. 020 7388 4097

Recommended reading

Metcalfe D D, Sampson H A and Simon R A, *Food Allergy: Adverse Reactions to Foods and Food Additives*, Blackwell Scientific Publications, Boston, 1991.

5.9 References

1 KJELLMAN N-I M, 'Adverse reactions to foods; management and prognosis', *Pediatr Allergy Immunol*, 1995 **6** (suppl 8) 54–8.
2 BOUSQUET J, BJORKSTEN B, BRUIJNZEEL-KOOMAN C A F M, HUGGETT A, ORTOLANI C, WARNER J O and SMITH M, 'Scientific criteria and the selection of allergenic foods for product labelling', *Allergy*, 1998 **53** 3–21.
3 TAYLOR S L and HEFLE S L, 'Food science perspective on food allergy', *Allergy*, 1998, **53** (suppl 46) 5–7.
4 DAVIS P J and WILLIAMS S C, 'Protein modification by thermal processing', *Allergy* 1998 **53** (suppl 46) 102–5.
5 GOLDSBOROUGH J and FRANCIS D E M, 'Dietary management'. In *The Second Fisons Food Allergy Workshop*, Medicine Publishing Foundation,

Oxford, 1983: 89–94.
6 BAHNA S L, 'Food challenges procedures in research and in clinical practice', *Pediatr Allergy Immunol*, 1995 **6** (suppl 8) 49–53.
7 WHARTON B and HIDE D W, 'The role of hypoallergenic formulae in cow's milk allergy and allergy prevention', *European Journal of Clinical Nutrition*, 1995 **49** (suppl 1) S1–S106.
8 RUGO E, WAHL R, and WAHN U, 'How allergenic are hypoallergenic infant formulae?' *Clin Exp Allergy*, 1992 **22**(6) 635–9.
9 ARSHAD S H, MATTHEWS S, GANT C and HIDE D W, 'Effect of food and house-dust mite allergen avoidance on development of allergic disorders in infancy', *Lancet*, 1992 **339** 1493–7.
10 SOGN D, 'Medication and their use in the treatment of adverse reaction to foods', *J Allergy Clin Immunol*, 1986 **78**(1) 238–43.
11 FISHER M, 'Treating anaphylaxis with sympathomimetic drugs', *B Med J*, 1992 **305** 1107–8.
12 EDWARDS A M, 'Diet and sodium cromoglycate'. In *The Second Fisons Food Allergy Workshop*, Medicine Publishing Foundation, Oxford, 1983: 95–7.
13 EWAN P W, 'ABC of allergies: anaphylaxis', *B Med J*, 1998 **316** 1442–5.
14 SAMPSON H A, 'Fatal food-induced anaphylaxis', *Allergy*, 1998 **53** (suppl 46) 125–30.
15 PATEL L, RADIVAN F S and DAVID T J, 'Management of anaphylactic reactions to food', *Arch Dis Child*, 1994 **71** 370–5.
16 ATHERTON D J, 'Dietary treatment in childhood atopic eczema'. In *The Second Fisons Food Allergy Workshop*, Medicine Publishing Foundation, Oxford, 1983: 109–10.
17 WARIN R P, 'Mode of action of some food additives in exacerbation of urticaria'. In *The Second Fisons Food Allergy Workshop*, Medicine Publishing Foundation, Oxford, 1983: 101.
18 SIMON R A, 'Update on sulphite sensitivity', *Allergy*, 1998 **53** (suppl 46) 78–9.
19 EGGER J, CARTER C, WILSON J, TURNER M W and SOOTHILL J F, 'Controlled trial of diet in migraine'. In *The Second Fisons Food Allergy Workshop*, Medicine Publishing Foundation, Oxford, 1983: 133–4.
20 MONERET-VAUTRIN D A, 'Food antigen and additives', *J Allergy Clin Immunol*, 1986 **78**(5) 1039–46.

6

Sources of information and labelling

F. Angus and J. Smith, Leatherhead Food Research Association

6.1 Introduction

Whilst food labelling on manufactured foods appears detailed, it still does not identify every allergenic component within the product. This is because current labelling regulations in some countries permit the manufacturer to use class or generic ingredient names, such as vegetable oil, and not always to have to list the individual components of compound ingredients if they are present in small quantities.

The increased awareness of food allergy and intolerance in recent years has led to considerable discussion regarding the labelling of allergens. In Europe these discussions are ongoing and as yet there is no mandatory requirement for the labelling of allergens. Some major companies in the UK have started to declare the presence of allergenic components on a voluntary basis, but this varies from company to company in terms of which minor ingredients are always identified.

Identifying problem-causing ingredients in foods can therefore be extremely difficult and it was for this simple reason that food intolerance databanks were set up – to assist intolerance sufferers in identifying foods that were safe to consume and so aid in the management of food intolerance.

There remain, however, very few food intolerance databanks in existence. The first to be established was the ALBA databank in the Netherlands, and this remains the largest and most comprehensive system in the world. The setting up of ALBA was followed by the UK Food Intolerance Databank and, since 1996, a number of smaller databanks have been set up in other parts of Europe. They are still the exception rather than the rule, however, and the fact remains that in many countries the intolerant are offered very little or no dietary support.

6.2 The UK Food Intolerance Databank

6.2.1 History

The problems faced by food intolerance sufferers when shopping have long been recognised in the UK. It was really in the 1960s that some dietitians started to compile lists of foods free from various ingredients for their patients. In 1976, this was partially systemised when certain dietetic departments in the UK undertook responsibility for compiling specific food lists, such as lactose-free, wheat-free, etc., and these lists were made available to the British Dietetic Association office. Realising the value of this information, in 1984 the BDA started to centralise the collection and dissemination of free-from information for use by its members.

During the same year, a report called 'Food Intolerance and Food Aversion' was published by the Royal College of Physicians (RCP) and the British Nutrition Foundation. This report made several recommendations relating to improving awareness, diagnosis and management of food intolerance. It also recommended that: 'The feasibility of setting up a central databank for food composition be examined. Products which are free from ingredients known to be responsible for intolerance should be registered in the databank, and doctors and dietitians should have access to it.'

The Food and Drink Federation (FDF) took up the recommendation and set up a working party with representatives from the Leatherhead Food RA, the British Dietetic Association, the British Nutrition Foundation, the Royal College of Physicians and the Institute of Food Research, Norwich. In 1995, this working party agreed that it would be feasible to set up a central databank of branded products 'free from' ingredients acknowledged as being responsible for food intolerance in a clinically significant number of people. It was decided that the databank would be used to produce lists of branded food, free from one or more of these ingredients. The lists were to be used to assist in the diagnosis and management of food intolerance.

The databank was developed at the Leatherhead Food RA and the scope of the databank was decided by the BDA and RCP. Concerns over self-diagnosis and the use of unsupervised diets, particularly for children, meant that the lists generated from the databank were designed to be used in association with appropriate dietary advice. The databank was finally launched in 1987 and the 'free-from lists' generated were made available to State Registered Dietitians.

6.2.2 Operation

The databank is now run as a collaborative venture between Leatherhead, the FDF and the BDA. Leatherhead maintains the databank by contacting food companies for information and producing the master lists.

The operation of the databank is simple. It is compiled from information submitted on Product Registration Forms, in hard copy or in disk format, which participating companies complete for each product they wish to enter.

Companies are supplied with detailed notes giving definitions of 'free from' for each ingredient and they indicate on the form whether their product is free from or not free from that specific ingredient. The information is entered onto the databank, categorised into product groups, which is used to subdivide the booklets, and returned to the company for checking. When all the information has been collected and sorted by the host computer, the various free-from lists are generated. The information is subdivided into over 20 food categories, such as meat products, confectionery, desserts, etc., which makes the products easier to find. These lists are forwarded annually to the BDA, which is responsible for printing and distribution of the lists to its members. Approximately 8000 booklets are distributed annually by the BDA; companies are asked to notify Leatherhead during the year of any product formulation changes that might affect the status of a product listed. Dietitians are notified of these mid-year changes via the BDA. Dietitians may also request specific combination lists of the ingredients covered by the databank for people with multiple food intolerances. These lists are generated to order by Leatherhead Food RA.

6.2.3 Ingredients covered

The UK Food Intolerance Databank currently covers the following ingredients and additives: milk, egg, wheat, soya, BHA and BHT, sulphur dioxide, benzoates and azo colours. The list is, however, under regular review. For example, in light of recent concerns about peanuts and peanut oil, the possibility of including information on peanuts has been discussed. The main problem in adding a peanut or nut category concerns the severity of reactions to minute quantities of peanut or nut and a belief by many food manufacturers therefore that this information should not be released to a third party. By restricting circulation of free-from peanut lists to named customers, a company is able to notify the individuals directly if there is a recipe change or contamination problem. Future developments planned for the databank, however, may make inclusion of nuts and other allergens in the Food Intolerance Databank possible.

6.2.4 Likely future developments

Discussions are currently underway regarding the future development of the databank. The difficulties in identifying allergenic ingredients from product labels have not been resolved and there is still likely to be an important future role for the databank, even if the labelling of allergens becomes mandatory in the EC (see Section 6.7.2).

The main change in the databank is likely to be in the scope of ingredients covered and in terms of access. There are discussions to broaden the number of ingredients covered and make access to the information wider and more straightforward. The potential for hosting the databank on the Internet is being explored.

6.3 The Dutch Food Intolerance Databank (ALBA)

ALBA is perhaps the most influential of the food intolerance databanks worldwide. It was established in 1982 by the Agricultural University of Wageningen and became operational in 1984. Since 1988, the databank has been hosted by a division of the government research organisation Netherlands Organisation for Applied Scientific Research (TNO), located in Zeist. ALBA currently holds data on around 500 brands and 11 000 products from 150 manufacturers and retail organisations, representing approximately 25–40% of the total Dutch manufactured food market. The 'free-from' booklets represent just one of the services offered by LIVO – the National Information Centre for Food Hypersensitivity, based in The Hague. As well as distributing over 12 000 'free-from' lists to consumers every year, LIVO provides a telephone enquiry service and produces general consumer information on food allergy and intolerance. ALBA distributes a further 2000–3000 special combination 'free-from' lists to consumers in the Netherlands. LIVO advises consumers on any changes to the ALBA lists via monthly newsletters.

The UK databank was based on the ALBA system and it therefore operates in a similar way. The main difference with regard to the management of the two databanks is that of funding. The ALBA databank and LIVO are largely funded by the Dutch Ministry of Health. Participating Dutch companies do not therefore have to pay a registration fee to list products on the databank, and a larger number of different 'free-from' lists can be produced. In addition, access to the ALBA lists is not restricted to dietitians, and they are available direct to consumers.

6.4 European food intolerance databanks

6.4.1 Background to the European Food Intolerance Databanks project (EFID)

The European Food Intolerance Databanks project developed as a result of the success of the food intolerance databanks in the Netherlands and the UK. In the early 1990s, the Leatherhead Food RA was aware that, despite the usefulness of food intolerance databanks in the management of food intolerance, the UK and Dutch databanks were the only ones in existence in Europe. After considerable consultation with experts, Leatherhead submitted a proposal to the Commission of the European Communities under the Agriculture and Agro-Industries Research programme for funding, to create a new network of food intolerance databanks across Europe. The project was accepted for funding.

6.4.2 Project timings and partners

Work commenced on the project in September 1993 and the project ran for a total of 36 months, ending in August 1996. Coordinated by Leatherhead Food

RA, the project had partners in nine European countries – Austria, Belgium, Denmark, France, Germany, Greece, Ireland, the Netherlands and Portugal. In order to ensure a broad representation of views in each participating country, a National Databank Team was established in each participating country. The team comprised a National Databank Coordinator, who represented the team on the EFID project, together with representatives from industry organisations, dietitians and key food manufacturers. The role of the National Databank Team was to assist the National Databank Coordinator in the development of the databank, and to adapt the general principles of the European team to reflect national characteristics. Whilst the UK and Dutch databanks provided the model on which each national databank was based in terms of broad concept and format of data, scope was provided for each databank to reflect the particular character and eating habits of the country.

6.4.3 Hurdles to overcome

There were many problems to overcome with this project. One of the first difficulties was agreeing definitions for tolerance levels of the ingredients and additives included on the EFID databanks. A sub-committee of medical and analytical experts was set up to discuss the sensitivity of the latest techniques to detect allergens in food and current thinking as regards the threshold for intolerance reactions for different ingredients. There was also a considerable amount of discussion regarding what the lists should be called in each country. Since the start of the UK and Dutch systems, the term 'free from X' had been used to name the booklets, but in some countries this terminology presented legal problems, which would be unacceptable to manufacturers wishing to contribute information on their products. A range of alternative terms was therefore devised in each country, which were acceptable to both the food industry and the legal advisers.

In addition, there was found to be a variation in the legal liability between the different EC countries if a product was incorrectly declared as 'free from'. In most countries, there is no obligation to inform consumers who suffer from food intolerance of the presence or absence of potential allergens in food, but if the manufacturer chooses to inform the consumer he will be liable for the information provided. The liability for injuries and negligence were the elements most likely to vary in different Member States.

There was also a significant increase in awareness of food allergy and intolerance over the course of the project. As a result, the Codex Alimentarius started discussions on the Proposed Draft Labelling of Allergens that can Cause Sensitivity (Alinorm 95/22). The prospect of potential mandatory labelling of allergens in the future delayed and, in some cases, led to a cessation in the establishment of food intolerance databanks in certain countries. This was due to a belief that there would be no need for the databanks if all the ingredient information was available to the consumer via the product label. This view was not supported by all the partners, however, and many believed that food intolerance databanks would have an important future role. This was largely because, firstly, they avoid the need to scrutinise the labels of every product when shopping, which can be extremely time-consuming and frustrating. Secondly, they are useful where a consumer may not be able to identify an ingredient on a label that they should avoid; for example, an individual allergic to milk may not realise that casein is a milk derivative. Finally, it is likely that mandatory labelling of allergens will be restricted to a core list of allergenic foods and will not cover all food intolerances.

6.4.4 European countries that successfully set up new databanks

Greece was the first partner of the project to set up its food intolerance databank, in August 1996. The databank is run by the Technological Educational Institution (TEI) of Thessaloniki and covers eight food additives and ingredients: milk, egg, gluten, wheat, soya, sulphur dioxide, benzoates and azo colours. The databank is updated annually and currently they have approximately 400 products listed from 25 companies.

The Technical University of Graz (Erzherzog Johann Universität) established the Austrian Food Intolerance Databank in October 1996 with the financial assistance of the Austrian Ministry of Science. This databank covers 11 ingredients and additives: milk protein, lactose, peanut, soya, wheat, gluten, egg, fish, benzoate, sulphur dioxide, and azo colours. The databank now has 19 contributors, with a total of 1600 products entered.

Two other participants in the project – Belgium and Denmark – launched their food intolerance databanks shortly after the end of the project contract.

6.5 Other international databanks

6.5.1 South Africa

The South African Food Intolerance Databank (FIDB) was initiated in 1990 by the Grocery Manufacturers Association in South Africa, which modelled the databank on the UK system. In 1995, the Association for Dietitians in South Africa (ADSA) took over responsibility for the project and subsequently produced *The South African Free From Handbook of Food Products*, a single book listing free-from information in tabular form. The project was supported by a large number of institutions in South Africa, including the Department of Health, Food Legislation Advisory Group, Consumer Services Board and two coeliac groups. Originally, access to the book had been restricted to medical professionals only, and, although the book was largely distributed by dietitians, it was later made available to the general public through bookshops. The databank covered ten ingredients and additives: milk, lactose, egg, soya, wheat, rye and gluten, benzoates, sulphur dioxide, BHA and BHT, glutamates and tartrazine. Companies were charged for entering products, and the book produced from the databank listed product information from 38 companies, including data from three major South African retailers. The databank was self-supporting, funded by a grant from the Dietetic Association, money from company registrations, advertising, and the sales of the book, of which around 500 copies of the first print run were sold. However, a lack of funding has meant that, as yet, new editions of the book have not been produced.

6.5.2 Developments and discussions in other countries

During the life of the EFID project, there was considerable interest in the project from all over the world. There was specific interest in setting up national food intolerance databanks in Hungary, New Zealand and the United States, but discussions on these potential new databanks are still underway.

6.6 Food labelling in Europe: an outline

The emphasis of European law at European Commission (EC) level is on product and consumer safety, together with ensuring the smooth operation of the internal market. The provision of consumer information, enabling consumers to make an informed choice concerning the foods they buy, is also one of the fundamental concepts of current EC law. Such horizontal provisions have generally replaced the concept of harmonisation by control of specific product standards.

Labelling requirements are detailed by Directive 79/112/EEC, as amended several times. The provisions of this Directive apply to most prepacked foods (the labelling of a number of products such as cocoa and chocolate products, certain sugars, honey and preserved milks is still controlled by vertical or product-specific standards, but revisions to these intended to simplify and streamline provisions on these product categories are currently under discussion). Foods prepacked for direct sale, i.e. prepacked on the premises for sale over a delicatessen counter or similar, and non-prepacked foods are not covered by the scope of the Directive; Member States of the European Union (EU) may establish their own rules in this area. Other key labelling directives are 89/396/EEC on lot marking and 90/496/EEC on nutritional labelling.

What are the requirements of Directive 79/112/EEC and what is their relevance to the allergen labelling issue? This Directive was developed using the principle of functional labelling, ensuring that consumers are presented with essential information as regards the nature of the product to ensure consumer safety and fair competition. Producers and manufacturers can give additional labelling information, provided this is accurate and does not mislead the consumer.

The mandatory requirements are as follows; exemptions from these are laid down in certain cases:

- Product description
- Ingredients list
- Date of minimum durability
- Net quantity
- Usage and storage conditions, if necessary
- Declaration of origin if its omission would be misleading to the consumer
- Name and address of manufacturer or packer, or of the seller in the EU
- Quantitative ingredient declaration (fully in force from 14 February 2000).

The most important of these requirements from an allergy point of view is the need to declare the ingredients.

Generic names

Ingredients are declared in descending order of weight. Generally, the name of the ingredient given in the list of ingredients must be that of the ingredient if it was being sold itself; however, certain ingredients may be declared using class or generic names. These include:

- All types of vegetable oil, except olive oil (class/generic name: Vegetable oil)
- All caseins and whey proteins (class/generic name: Milk proteins)
- All types of starch (class/generic name: Starch)

Therefore, in certain cases, it is possible for a consumer not to be aware of the exact ingredients used in product manufacture, as the exact name of the ingredient used does not have to be given in the list of ingredients.

Compound ingredients – the 25% rule

Under this Directive, the components of a compound ingredient need not be declared in the ingredients list of a prepacked food if that compound ingredient is present in a quantity of less than 25%, other than any additives present having a direct technological effect in the final food. Under this rule, therefore, certain ingredients that have potential allergenic reactions, but are present only in small quantities, need not be specifically declared in the list of ingredients. Consumers who may be at risk are therefore not aware of their presence.

Processing aids

Additives present in a final food by means of carry-over from an ingredient used in the final product preparation that do not have a technological effect in the final food also need not be declared in the list of ingredients. In addition, processing aids need not be declared. It is possible that low levels of an additive causing hypersensitivity reaction in certain sectors of the population may be present without consumers being aware of it.

Declaration of the source of gluten

The most recent amendment to 79/112/EEC, Directive 97/4/EC, introduced a requirement that is of direct relevance to the allergy question. Put forward by the

European Parliament as an amendment to the proposed text that was in circulation, the requirement is now in place that if the generic name 'modified starch' or 'starch' is used in the ingredients list, this must be accompanied by an indication of the specific vegetable origin of the starch or modified starch, where this may contain gluten. The amending Directive is now in force. National implementation is now underway, with Member States having until 14 February 2000 to prohibit products not complying.

6.7 Current and proposed labelling requirements for ingredients causing hypersensitivity

6.7.1 Codex

Among the matters for adoption by the 23rd session of the Codex Alimentarius Commission in June 1999 is the Codex Alimentarius Commission Draft Recommendations for the Labelling of Foods that can Cause Hypersensitivity (Draft Amendment to the General Standard for the Labelling of Prepackaged Foods). As hypersensitivity to foods is an international issue, agreement on Codex provisions in this respect would form a basis for the application of relevant labelling rules in many countries whose national legislation is Codex-based. This Draft Amendment includes compounds triggering both food allergies and intolerances, owing to their importance from a public health point of view, and the list included covers both food groups and individual foods, on the basis of the recommendations of the FAO Technical Consultation on Food Allergies.

So what does the draft amendment, Alinorm 99/22, Appendix III, now at Step 8 of the Procedure, require? It requires that the following foods and ingredients known to cause hypersensitivity are always declared as such:

- Cereals containing gluten: i.e. wheat, rye, barley, oats, spelt or their hybridised strains and products of these
- Crustacea and products of these
- Eggs and egg products
- Fish and fish products
- Peanuts, soyabeans and products of these
- Milk and milk products (including lactose)
- Tree nuts and nut products
- Sulphite in concentrations of 10 mg/kg or above

Class names that may be used in certain cases would not be authorised for such ingredients listed above unless a general class name would be more informative. Exemptions from declaration for processing aids and additives carried over into foods at a level less than that required to achieve a technological function would not be valid for processing aids or additives included in the above list.

One of the key issues is what substances are included on the list and how this list should be amended in future; the Joint Expert Committee on Food Additives

(JECFA) has indicated that it would be willing to consider further evidence on foods that may cause hypersensitivity, so providing a scientific basis and criterion for the inclusion of foods on the list. There is still some concern that some of the categories are too broad, for example 'milk and milk products'; for 'soyabean, peanuts and their products' the protein fraction may be allergenic but there was no evidence to suggest that refined or heat-treated oils have the same effect. It is felt that these issues should not delay the establishment of the list.

The 25% rule

The issue of the 25% rule has been separated from that of establishing the list of substances causing hypersensitivity for labelling purposes. The issue under debate is whether or not the 25% rule should be modified so that ingredients of a compound food are exempt from declaration only if the compound food is present at less than, say, 5%, or whether the exemption should be removed entirely, so leading to longer, more complex ingredients lists. Would reduction of the percentage be of value in this instance as many substances causing hypersensitivity act at very low levels? There would, therefore, be little scientific basis for changing the 25% limit on these grounds alone. The labelling of many other ingredients would be affected by changes in the 25% rule; therefore, this part of the draft has been returned to Step 6 for further discussion.

6.7.2 The EC situation

Currently, under EC labelling legislation, there is no need to make a specific declaration in respect of the presence of potential allergens, although product liability laws might influence a decision to make such a declaration. However, the need for such information to be covered by food labelling legislation, in line with current Commission intention, is now widely recognised.

A further amendment to Directive 79/112/EEC is already under discussion, which would concern the labelling of potential allergens when present as ingredients. In the opinion of the European Commission, the consumer does not receive detailed information about the exact composition of the foodstuff he or she is buying, owing to the compound ingredient provisions, although he or she can still make an informed choice. However, the lack of such detailed information can be problematic to those with allergies or intolerance to certain substances, who need as much information about the product as possible. Although it is recognised that labelling for consumers in general must not be considered as the only source of information available, as the medical establishment is key in this respect, it is advisable to assist those with allergies or intolerance as much as possible by making more comprehensive information about the composition of products available to them. Therefore, it is considered necessary that certain substances recognised scientifically as being the source of allergies or intolerance be included in a list of ingredients and not qualify as exceptions under the general labelling Directive. The Commission recognises that Member States can take their own action concerning foods sold in bulk or foods served in catering establishments. As the Directive does not apply to non-prepacked foods, the introduction of the amending Directive in due course will still mean that consumers need not be informed under EC law if non-prepacked foods contain potential allergens.

The document currently under discussion is the draft proposal for a European Parliament and Council Directive amending Directive 79/112/EEC on the approximation of the laws of the Member States relating to the labelling, presentation and advertising of foodstuffs, Document III/5909/97, dated 6 January 1998. Under this proposal, the following ingredients will be required to be declared on a label, whatever their level in the food:

- Cereals containing gluten and products of these
- Crustaceans and products of these
- Eggs and egg products
- Fish and fish products
- Peanuts and products of these
- Soyabeans and products of these
- Milk and milk products (lactose included)
- Tree nuts and nut products
- Sesame seeds
- Sulphite in concentrations of at least 10 mg/kg.

Gluten-free foods

Under EC Directive 89/398/EEC, as amended, foods for particular nutritional uses, or PARNUTs foods, are defined as foods that, owing to their special composition or manufacturing process, are clearly distinguishable from foodstuffs for normal consumption, which are suitable for their claimed nutritional purpose and which are marketed in such a way as to indicate suitability. Currently, there are nine categories of PARNUTs foods listed in the annex to this Directive, one of which is gluten-free foods. The intention was originally to establish detailed compositional and labelling requirements on each of these categories of foods; however, only certain categories have so far been regulated in this way and there is ongoing discussion as to whether gluten-free foods and the other remaining categories should be included under the scope of this Directive at all. If a decision is taken to remove gluten-free foods from a PARNUTs classification, it would probably be considered that labelling provisions would be an adequate means of control.

Therefore, it is up to the Member States at this time to determine to what extent labelling of allergens and ingredients likely to cause hypersensitivity is required.

UK labelling

There are, currently, no specific provisions under UK legislation concerning the labelling of potential food allergens. The Food Safety Act 1990 requires that food be of the nature, substance or quality demanded by the purchaser and comply with food safety requirements. In addition, labelling must not be misleading to the consumer.

The Food Labelling Regulations 1996, SI 1996 No. 1499, control the mandatory labelling requirements for foods. These implement the general provisions of 79/112/EEC, but there are no specific references to the labelling of allergen ingredients. However, the Ministry of Agriculture, Fisheries and Food (MAFF) drew attention to the labelling of food containing nuts in draft guidance notes to the 1996 Regulations; the draft notes recommended that if the presence of nuts was not clear from the product name, the ingredients list or the way in which the food was presented, an appropriate warning should be given on the label, for example a 'contains nuts' declaration placed prominently on the label or a *new recipe* declaration, which would warn the consumer to look at the ingredients list more closely. These comments never reached the final published version of the guidance notes, possibly owing to developments at EC level being awaited. In the UK at this time, some manufacturers choose to give a declaration such as 'contains nuts' or 'may contain traces of peanuts'; provided the general requirements of the Food Safety Act and the Food Labelling Regulations are not contravened, such labelling can be acceptable.

Germany

There are no guidelines or regulations on the labelling of foodstuffs that may cause allergens, or substances that may cause a hypersensitive reaction. The Federal Ministry of Health has indicated that it welcomes the efforts now underway to deal with this issue, but does not plan at present to draw up any specific provisions. As in other countries, statements such as 'does not contain nuts' could be a problem under product liability laws.

Sweden

In contrast, Sweden is an EU Member State with detailed guidelines on allergen labelling. Under the Swedish labelling regulations, the compound ingredient rule in the EC general labelling Directive 79/112/EEC is applicable. However, if the compound ingredient is listed only with its name, it is desirable that ingredients that may cause hypersensitivity reactions are always stated in the list of ingredients. Examples of such ingredients given in the 1997 Guidelines on the Labelling of Foods are gluten-containing grain, eggs, milk, fish, nuts, leguminous plants (e.g. soyabeans, peanuts and peas), and sulphite. For example, the labelling could be given in the form 'margarine (contains milk)' or 'mayonnaise (with eggs)'. Although some consumers may be aware that these foods are likely to contain such components, others may be less so. It is also recommended that the same guidance is applied to additives that may cause hypersensitivity reactions, i.e. they should always be declared in the list of ingredients – for example antioxidants, colours and preservatives. Although additives are declared in compound foods if they have a technological effect in the final foods, if they are present only by carry-over and are not technologically effective, then their declaration is not required under current laws.

An amendment to the Swedish labelling laws, dated 1995, included changes to the guidelines on the use of certain claims. It is now stipulated that labelling may not contain expressions, symbols or other information suggesting that ordinary foods are intended for particular nutritional purposes. 'Naturally gluten-free', 'free from milk', 'without soya' and 'suitable for allergenics' are given as specific examples of expressions that should not be used on ordinary foods. Symbols with such meaning should also not be used.

Finland

The Finnish labelling regulations include certain requirements for the declaration of potentially allergenic ingredients, as part of the provisions regarding the declaration of compound ingredients. A compound ingredient may be declared by its own name, provided that the list of its own ingredients and additives immediately follows this name. If the compound ingredient represents less than 25% of the final product, of the ingredients used, at least the 'active' additives and those ingredients that can produce symptoms of hypersensitivity in an individual using the foodstuff must be declared. The following at least must be declared in this respect: peas, fish, eggs, milk, soyabeans and crustaceans, and products manufactured from them; peanuts, almonds and nuts; and oats, barley, rye and wheat.

The problem of allergenic and hypersensitivity reactions is an international one, and this is reflected in provisions in certain other major international markets.

6.7.3 Australia

Under the Australian Food Standards Code, Standard A1, compound ingredient provisions are detailed whereby, in common with the EC, if an ingredient contributes less than 250 g/kg (25%) of a food, food additives are the only components that must be declared. In contrast to EC law, unless specifically required, if an ingredient contributes less than 100 g/kg (10%) of a food, the components of the food need not be declared. However, the Standard does require that the presence of peanuts must always be declared in a food. Standard K2 on honey and related products requires that labels on or attached to pollen products must declare, in standard type of 3 mm, the statement 'THIS PRODUCT MAY CAUSE SEVERE ALLERGIC REACTIONS'. The same standard requires the following statement on packages containing royal jelly, or to foods containing royal jelly, immediately following the name of the food, in type of 3 mm: 'THIS PRODUCT CONTAINS ROYAL JELLY, WHICH HAS BEEN REPORTED TO CAUSE SEVERE ALLERGIC REACTIONS AND, IN RARE CASES, FATALITIES, ESPECIALLY IN ASTHMA AND ALLERGY SUFFERERS'.

As part of the review of the Food Standards Code, the Australia New Zealand Standards Authority has proposed to revise current Australian and New Zealand regulations for specific labelling statements such as warning statements and labelling of foods that may cause severe adverse reactions. According to the

Australian proposals, a food or food additive to be included in a list of components that must be declared must be recognised by medical experts as a frequent cause of severe systematic reactions resulting in significant morbidity or mortality. The list of foods and ingredients proposed is very similar to that put forward by the EC.

6.7.4 United States

Of particular interest is the Food and Drug Administration (FDA) position in this respect. In June 1996, the Center for Food Safety and Applied Nutrition issued a Notice to Manufacturers concerning the Label Declaration of Allergenic Substances in Foods.

The Food, Drug and Cosmetic Act requires, in virtually all cases, a complete ingredients listing on foods. Two exemptions to this are that spices, flavourings and colourings may be declared collectively under the Act, without each individual one having to be specifically named; also, incidental additives, such as processing aids, that are present in foods at an insignificant level and do not have a technical or functional effect in the final food, need not be declared, under Title 21 of the Code of Federal Regulations. The FDA, in this Notice, stressed to manufacturers that the exemption applied only when the incidental additive is present at an insignificant level *and* it must not have any technological effect in the final product. An example is quoted of egg white as a binder in breading on a breaded fish product; the egg white is not incidental as it is acting in the final food, so should be declared. Owing to the low levels of ingredients concerned with allergens, the FDA is considering whether it is necessary to clarify the regulations to ensure that manufacturers fully understand the circumstances in which allergenic food ingredients must be declared and to ensure that sensitive individuals are protected by appropriate labelling. The FDA is also open to comment on how the problem of potential allergens in additives should be handled. It may consider it necessary to introduce rule-making for the labelling of allergenic ingredients.

While assessing the situation, the FDA, in the Notice, requests manufacturers to examine their product formulations for ingredients and processing aids containing known allergens that are currently exempted from declaration as incidental additives and to declare these in the ingredients statement. Where appropriate, the name of the ingredient may be accompanied by a parenthetical statement for clarity, for example '(processing aid)'. It is felt that allergenic ingredients in an additive could be declared in the correct position in the list (owing to their low levels, usually at the end) and other non-allergenic ingredients would continue to be exempt.

Examples of foods that are among the most commonly known to cause serious allergenic responses are, according to the FDA, milk, eggs, fish, crustacea, molluscs, tree nuts, wheat and legumes (in particular soyabeans and peanuts). The FDA advises that the issue of declaring allergenic ingredients in food is being discussed on an international level – a move it welcomes.

Another area of concern to the FDA is cross-contamination so as to cause inadvertent addition to or introduction of an allergenic ingredient into a product where it would not normally be found. For example, a product without peanuts could end up containing peanut traces. The FDA feels that labelling such as 'may contain peanuts' should not be used as a substitute for Good Manufacturing Practice (GMP); manufacturers are urged to take all steps to eliminate such contamination and ensure the absence of the allergenic food or ingredient. The FDA is considering options for providing consumers with further information in this respect and how this issue should be addressed.

6.7.5 Canada

The Canadian Food and Drug Regulations require the specific declaration of peanut oil, hydrogenated or partially hydrogenated peanut oil and modified peanut oil, wherever they are present, whether they are added as such or are components of ingredients.

In 1998, the labelling of foodstuffs that cause severe reactions in certain individuals was the subject of a review by a joint committee from the Canadian Food Inspection Agency and Health Canada. The Canadian authorities have recognised the importance to consumers of labelling those ingredients most likely to elicit adverse and abnormal reactions. With this in mind, they have put forward recommendations that the list of foods that must be specifically listed in foodstuffs is increased to include peanuts, named tree nuts, sesame seeds, milk, eggs, fish, named crustaceans and shellfish, soya, wheat and sulphites. This list has subsequently been endorsed by the Canadian Society of Allergy and Clinical Immunology and other interested parties, and the publication of a proposed regulation regarding the labelling of sulphites is expected in the near future.

The joint committee has also reviewed the present labelling requirements for foods derived from plants and recommends that the plant source of hydrolysed plant proteins, starches, modified starches and lecithin should be included, in their common name, in the list of ingredients. Such identification would prevent the unnecessary dietary restriction of individuals who are sensitive to specific plant species.

The federal government has also developed a precautionary labelling policy, to allow manufacturers voluntarily to label products that may inadvertently contain substances capable of causing severe adverse reactions.

6.8 Future labelling trends

Following on from the publication of the Green Paper on Food Law by the European Commission, which recognised the need to review and consider whether or not current labelling requirements best satisfy the information needs of consumers, a number of aspects of food labelling are under review. The provision of information to consumers is considered to be of paramount importance, in order to allow them to make informed decisions, and is becoming increasingly recognised as essential where a public health risk is concerned. It has not, therefore, been surprising that the labelling of potentially allergenic compounds in foods has become a key issue, particularly as product liability laws will also need to be considered in the event of a consumer having a reaction from an unlabelled product. Although a number of countries have their own provisions in place or at the proposal stage, there is considerable interest in the Codex and EC proposals in this respect. Owing to the reference role of Codex standards in World Trade Organisation agreements and possible trade disputes, many countries will undoubtedly focus on Codex provisions once agreed. Adoption of an amending Directive at EC level will have direct implications for each of the EU Member States and for those wishing to become part of the expanded Union in the future.

Two issues are apparent and of concern in proposals that have been published so far. The first aspect is the need for any list of potentially allergenic or hypersensitive ingredients to be based on scientific evidence and to be established in such a way that it can readily be changed if required, for example by addition of a new ingredient. The second issue concerns that of non-prepacked foods and foods prepacked for direct sale. Such foods are often outside the scope of labelling regulations (for example Directive 79/112/EEC, as amended), but may be the more significant potential problem in this area. Possibly the use of tickets or notices for labelling is an option, and staff knowledge and training are of paramount importance. Currently, it is up to countries' individual regulatory requirements whether or not such labelling is required. As food labelling regulations continue to reflect consumer safety and information needs, it is inevitable that more and more countries will include provisions on the labelling of potentially allergenic substances in their national law. The importance of Codex and EC provisions in this area cannot be overestimated as, once agreed, they will play a significant role in minimising the potential for the labelling of ingredients causing hypersensitivity to form trade barriers.

6.9 Sources of further information and advice

6.9.1 Consumers

'Free-from' lists from supermarkets and companies

'Free-from' lists covering own-label products are available from Waitrose, Sainsbury, Tesco and Safeway. These lists cover the standard ingredients covered by the UK Food Intolerance Databank, plus peanut and nuts. Asda contributes data on its own-label products to the UK Food Intolerance Databank.

Some of the major food companies also provide 'free-from' information on their products direct to customers. Most of the major food companies will provide peanut and nut-free information to their customers directly.

Coeliac Society gluten-free lists
The UK Coeliac Society has been providing information and advice to its members for many years. The Society compiles its own 'List of Gluten-Free Manufactured Products' annually and this is available from its head office in High Wycombe.

Self-help groups
There are a number of self-help groups that offer information and advice to sufferers. In the UK, the British Allergy Foundation and the Anaphylaxis Campaign are among the most important. The British Allergy Foundation provides information, advice and support to allergy sufferers, including a helpline, a regular newsletter and leaflets. The Anaphylaxis Campaign works to raise general awareness of severe food allergies and provides general advice and a video on anaphylaxis, as well as producing a quarterly newsletter.

In the United States, the Food Allergy Network provides a wide range of assistance for food allergy sufferers, including general advice, product alert information, plus a video and a bi-monthly newsletter.

Company information
Individual companies may have their own policies regarding labelling of potential allergens at this time, and information may be available to the public by means of direct communications or by printed material.

6.9.2 Information for industry

Publications, CD-ROMs and training materials
There is an increasing number of publications and CD-ROMs that act as reference sources for UK, EC or international food legislation. These include the Leatherhead Food RA publications *Guide to Food Regulations in the UK*, *EC Legislation* and *International Food Legislation Manual*, all available in hard copy or in CD format. Traditionally recognised publications such as *Bell and O'Keeffe's Sale of Food and Drugs* and *Butterworths* in the UK are valuable reference sources. The Eurolaw CD-ROM enables ready reference to proposed and actual legislation.

In 1999, the Food and Drink Federation in the UK published *Food Allergens Advice Notes*. These summarise the current legal position with regard to labelling of allergens and liability issues and provide advice regarding handling allergenic ingredients in the factory and use of defensive labelling.

In 1993, the Food and Consumer Products Manufacturers of Canada (FCPMC) in association with Health Canada and the Allergy and Asthma Information Association produced an industry training programme called 'Allergy Beware'. The purpose of the programme, which includes a teaching guide, a video and a factory audit checklist, is to raise awareness about anaphylaxis in the food industry.

The Food Allergy Research and Resource Program (FARRP) at the University of Nebraska provides analytical services, information and training to the US food industry on issues related to food allergy.

Internet

Many countries now publish information about actual and proposed legislation on their official websites; some also publish the text of official journals, even if for only a limited period, enabling the Internet searcher to keep up to date with developments in the country concerned.

Many of the organisations listed above under 'Publications, CD-ROMs and training materials' have websites on the Internet, which provide background information on food allergy as well as details on their services to industry.

7

Analytical techniques for detecting food allergens

S. Kilburn, Queen Alexander Hospital, Cosham

7.1 Introduction

Analytical techniques are available to detect very low concentrations of allergenic proteins within complex mixtures. These have potential for use in the food industry in a number of ways:

- Quality control of food manufacturing processes – detection of accidental cross- or carry-over contamination due to shared areas, staff or equipment.
- Confirmation of accidental exposure to an individual.
- Detection of residual allergenicity following processing to reduce biological activity.

There are several considerations to be made when designing or implementing analytical assays:

- Nature of the allergenic molecules
- Detection limits required
- Sensitivity and specificity of the assay
- Sampling and extraction procedure.

The aim of this chapter is to outline the methods available, and to outline the limitations and benefits of such assays, but not to give a step-by-step guide to bench-top techniques. The chapter begins with a brief description of the type of molecules that we need to detect. The detection limits required have been indicated in Chapters 3 and 4 where the extreme sensitivity of some individuals to minute quantities of proteins has been described and the threshold doses discussed. It is apparent that for certain foods, such as nuts and peanuts, as little as 45 mg (Hourihane *et al.* 1997) must be detected in a meal in order to avoid risk of anaphylaxis. This level of sensitivity must therefore be achieved with the assays. Protocols for sampling are given in Chapters 3 and 4. Sampling procedure could greatly compromise the efficiency of detection, especially if the contamination is likely to be particulate and intermittent. Enzyme-linked immunosorbent assay (ELISA) is an aqueous system, and allergens must be in aqueous form for analysis. Poor extraction and recovery of the allergen could compromise an otherwise adequate assay. All these factors and steps must be considered when implementing allergen detection assays.

7.2 The physical and chemical nature of food allergens

7.2.1 Foods that commonly cause allergy

Foods that can give rise to allergic reactions in susceptible individuals appear to be diverse in nature. However, although reactions to many different foods have been described in individual case reports, the list of *common* causal agents is relatively short. This has led researchers to postulate that there may be certain features characteristic of food allergens. Common causes of allergy are milk, egg, peanut, tree nuts, fish, shellfish, soy and citrus fruits for populations in the UK and the USA. The list can vary for different countries; for example, Mediterranean countries such as Italy have a high incidence of sensitivity to olives, and in Japan even sensitivity to birds' nest soup has been described.

To be capable of inducing an allergic reaction a food must contain substances that are immunogenic, and give rise to allergic sensitisation. This results in the production of IgE antibodies in preference to IgG and T cells of the Th2 phenotype rather than the Th1 phenotype. On subsequent exposures the molecule must be able to cross the mucosal barrier and cross-link IgE on effector cells, causing degranulation and release of the chemical messengers that produce allergic symptoms. The molecule must therefore bear more than one IgE binding site. The majority of described allergens are protein in nature, though carbohydrates/sugar moieties may also cause symptoms as they certainly bind IgE. Carbohydrate epitopes may be responsible for cross-reactivity between plant species (Blanco *et al.* 1999, Caballero and Martin-Esteban 1998). Lipids (fats and oils) do not provoke a specific immune response and so are not causal for allergic reactions. Current allergen detection techniques and diagnostic assays focus on the protein components.

7.2.2 The basic structure of proteins

The building blocks of proteins are amino acids, bound together in a linear fashion by covalent peptide bonds. Each protein has a precise length and amino acid sequence dictated or transcribed by messenger RNA that in turn is translated from the DNA. Once made the protein may be modified, or chopped into smaller pieces, or carbohydrate, lipid or phosphate moieties may be added by the action of enzymes within a plant or animal cell. The linear sequence of amino acids is termed the *primary structure* of the protein. Proteins are rarely linear in the native form but form distinctive three-dimensional structures. This is due to chemical interactions between amino acids in close proximity, causing the chain to form twists that force it into spirals, termed alpha-helixes, and sharp bends resulting in so-called beta-pleated sheets. These basic forms are part of the *secondary structure* of the protein. The arrangement of these secondary structures in relation to each other gives rise to the *tertiary structure*. This results from non-covalent interactions between the different regions of the same protein or polypeptide molecule. In addition many functional proteins consist of aggregates of two or more polypeptide chains, that are homogeneous or heterogeneous. This is termed the *quaternary structure*. The three-dimensional shape and chemical nature of the amino acid backbone and additional groups contribute to the functional and antibody-binding properties of the molecule.

7.2.3 Molecular characteristics of common allergens

The majority of allergens described are protein in nature with or without carbohydrate moieties (glycosylated), with a molecular weight ranging between 10–100 kDa. Most proteins in foods can be immunogenic and provoke production of specific antibody, mainly IgG, in individuals with or without an atopic tendency. Only a limited range of proteins is commonly associated with the production of IgE in the atopic individual, and is considered allergenic. Protein molecules that initiate immune responses are commonly over 7000 daltons in size (Roitt *et al.* 1998). No common molecular motif for allergens has been described, but they do have some properties in common. Allergens, particularly those that lead to persistent allergies, are thought to be resistant to digestion (Astwood *et al.* 1996, Becker 1997), the rationale being that this results in persistence in the body and stimulation of the immune system. There are certain fruit allergens, which may be unstable, even being degraded by enzymes released in the fruit by crushing (Bjorksten *et al.* 1980). Many allergens have enzymatic ability (Bufe 1998) so function in addition to stability may be related to allergenicity. Commonly a food will contain more than one allergenic protein, such as beta-lactoglobulin, lactoferrin and the caseins of cow's milk, and ovomucoid, ovalbumin and lysozyme of egg, indicating that the context as well as molecular structure must be important.

7.2.4 Techniques for identifying allergens and quantifying allergenicity

A number of techniques have been used to identify allergenic proteins, most being based on the principle of:

- Solubilising/extracting proteins
- Isolation of protein fractions
- Determining IgE binding ability of each fraction
- Characterisation of the protein/glycoprotein and larger-scale purification.

The techniques most often used in the current literature for allergen identification are:

- Separation on a gel such as SDS PAGE followed by Western blotting and immuno-labelling
- Separation by chromatography (often High Performance Liquid Chromatography) followed by ELISA.

Using ELISA or Western blotting, quantitative or semi-quantitative data on the binding of serum IgE to specific proteins can be calculated for individual patients. Generalisations on allergenicity of specific proteins in a food are made by assessing the proportion of affected individuals that have elevated IgE to that protein. These methods cannot predict the degree of symptoms that may be produced on exposure to each individual protein or the outcome of introducing novel foods into a community.

7.3 Principles of food allergen detection techniques

The choice of assay has a great effect on the sensitivity and specificity. There are some foods where the sensitivity is paramount whereas specificity is not, such as detection of protein in oils extracted from allergenic seeds. In most other situations proteins will normally be present in the food and specificity without sacrificing sensitivity is required.

7.3.1 Protein detection

A number of assays have been developed to quantitate proteins in solution. All are susceptible to interference by other compounds that may be present. The Bradford method is widely used, but the BCA method is more robust. However, the latter is sensitive to interference from reducing sugars. These assays give us an approximation of the quantity of protein present but not whether these proteins are allergens or not. They are, however, useful for the estimation of residual protein in, for example, oils extracted from seeds where the source material is known to be allergenic.

The Bradford Method (Bradford 1976)

This assay makes use of the acidic dye, Coomassie Brilliant Blue G-250, which binds to any basic and aromatic amino acids present on the polypeptide molecule. This changes the colour of the dye from brownish (absorbance at 465 nm) to blue (absorbance at 595 nm). The colour change is recorded using a spectrophotometer at wavelength 595 nm and the results are read from a standard curve generated from a protein of known concentration. A good description of the technique is provided in Rosenberg (1996), and the detection limit of the assay is approximately 200–1400 μg/ml. Reagents are available from Sigma-Aldrich.

The Bicinhoninic Acid Method (Smith et al. *1985)*
When a protein is placed in an alkaline system containing Cu^{2+}, a coloured complex forms between the peptide bonds of the protein and the copper atoms. Bicinhoninic acid forms a complex with cuprous ion (Cu^{1+}) in an alkaline environment, resulting in a stable, highly coloured chromophore with an absorbance maximum at 562 nm. The sensitivity of the assay is approximately 0.5–10 μg/ml. See Rosenberg (1996) for a description of the method.

7.3.2 Detection of specific proteins – the immunoassay

The most commonly used technique for quantification of allergenic or antigenic substances is the enzyme-linked immunosorbent assay (ELISA). ELISA has the advantage over radioimmunoassay (RIA) of being more cost-effective and, with modern techniques, not compromising sensitivity. The specificity of all immunoassays is in part dependent on the efficiency of the capture and detector antibodies. Once optimised and standardised the ELISA is relatively economical, and large numbers of samples can be analysed on each test run. The assay is carried out in standard plastic 96 well plates designed for use in ELISA. The wide use of such plates has led to a variety of plate washing and reading systems being available. The sensitivity of the antibodies in forming a complex with the protein is paramount for the sensitivity and specificity of the assay. The sensitivity of the basic assay may be further increased by using indirect labelling or amplification techniques. In non-competitive assays all the constituents are in excess, apart from the protein to be detected. The optimum quantities of each constituent are determined by preliminary experiments. An alternative is the inhibition ELISA, also highly sensitive, but this technique is susceptible to non-specific interactions.

Sandwich enzyme-linked immunosorbent assay
The two-site or sandwich ELISA is ideal for detecting proteins in complex mixtures (Fig. 7.1). The antibodies may be from monoclonal or polyclonal origin. Monoclonal antibodies are most often used for capture, since polyclonal antibodies with wider specificity may theoretically mask the binding site for the monoclonal antibody. It is important that the capture antibody does not interact directly with any of the subsequent assay stages, or vice versa, as this leads to abnormally high background values that reduce sensitivity. In the majority of assays antibodies from different animal species are used to avoid this. An assay may utilise polyclonal antibodies both capture and detector stages. Monoclonal antibodies may be used for both capture and detector stages. In this situation the assay designer usually ensures that the capture and detector antibodies are directed against different parts of the molecule (so-called two-site assays) to avoid competition or interference between the antibodies. The example given in Fig. 7.1 is a direct assay where the enzyme label is directly conjugated to the antibody. Amplification steps are described below.

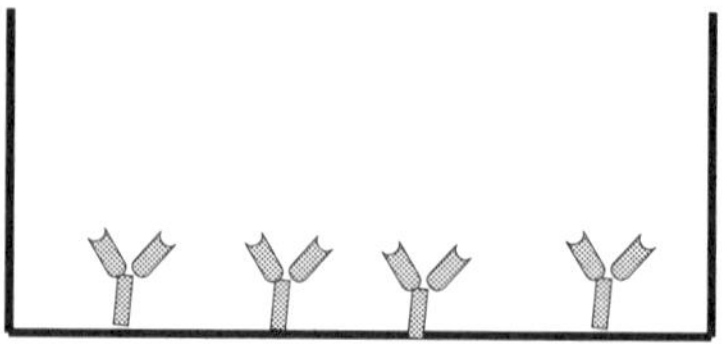

(a) Protein, in this case 'capture' antibody, is incubated in the well, usually of alkaline pH, and small quantities become absorbed onto the plastic surface.

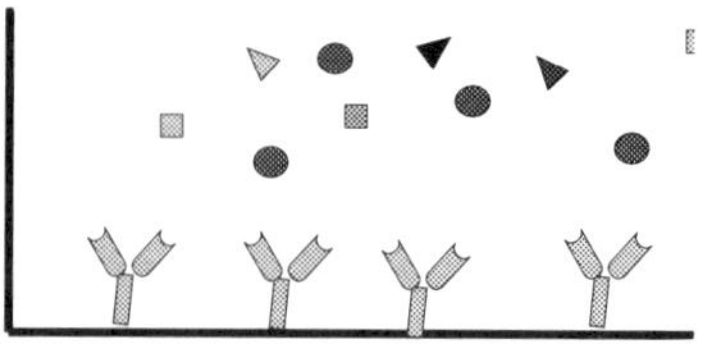

(b) The protein solution to be tested is added to the plate at the appropriate dilution in duplicate or triplicate.

(c) The capture antibody binds the specific protein, anchoring it to the plate surface. Any unbound proteins will be washed off the plate and discarded.

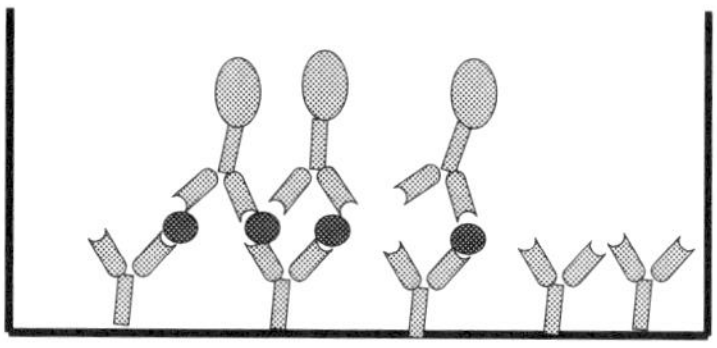

(d) A 'detector' antibody specific for the protein is then used to link an enzyme label to the protein.

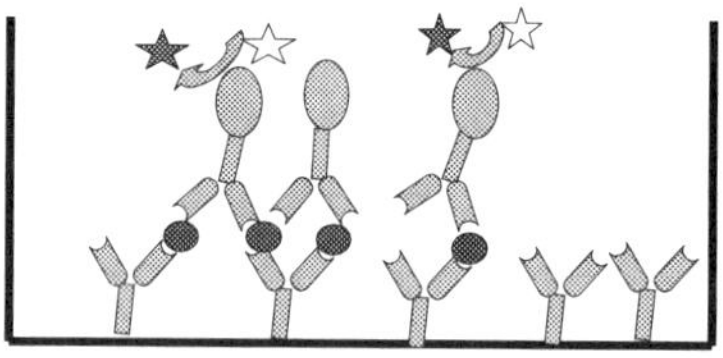

(e) The enzyme then acts on a colourless substrate to form a coloured product that is quantitated using a specialised optical density plate reader. The generation of standard curves using known antigen concentrations allows for the accurate estimation in weight/volume.

Fig. 7.1 The enzyme-linked immunosorbent assay (ELISA) – antigen detection by direct sandwich.

Choice of antibody

The assay specificity and to some extent its sensitivity are primarily linked to the efficiency of the antibodies used. Polyclonal antibodies are cheaper to prepare than monoclonal; both types of antibody should be purified to reduce non-specific interactions, and in the case of polyclonal antibodies they may also need to be affinity purified. Polyclonal antibody preparations contain a heterogeneous mix of antibodies directed against any number of epitopes on the protein surface. This gives the assay an advantage if food processing results in some epitopes being denatured or masked, but increases the likelihood of non-specific interactions between the antibodies and unrelated proteins. Monoclonal antibody preparations have a specificity directed against one epitope; this may increase specificity greatly if the epitope is not present on unrelated proteins. The assay may prove to be less versatile if the epitope is more susceptible to denaturing than the allergenic epitopes.

Amplification systems

The detector antibody may be conjugated directly to the enzyme label – direct assay. Or various amplification steps may be used:

- Conjugation of the detector antibody to biotin, a compound that binds to avidin with a great affinity and specificity, thus amplifying the signal – indirect biotinylated assay
- The detector antibody, for example IgG from rabbit, may be unlabelled, and subsequently detected using a third enzyme-labelled antibody specific for the detector antibody – indirect assay.

See Fig. 7.2.

Substrates

Substrates of choice in traditional assays gave a coloured product or chromogen, but more recently fluorescent or luminescent substrates are being used. The

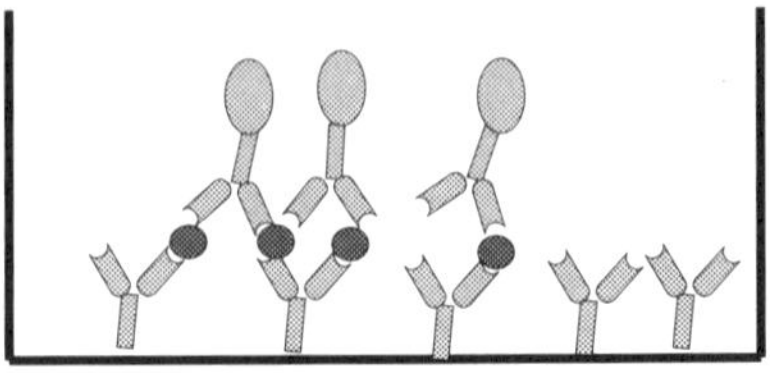

(a) In the direct ELISA the primary antibody is conjugated to the label.

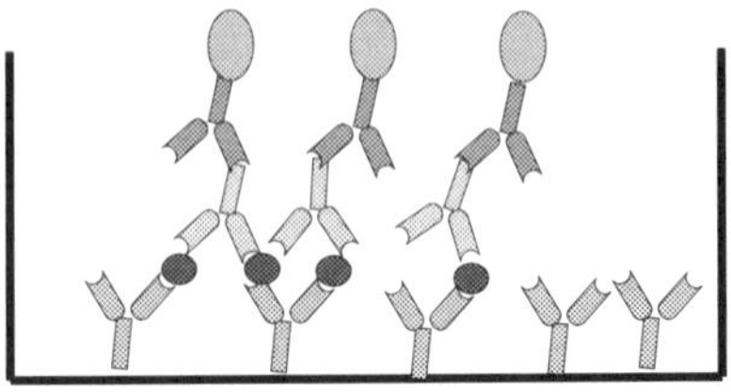

(b) In the indirect ELISA the primary antibody is unlabelled and a secondary antibody carries the label. There must be no interaction between the capture and secondary antibodies.

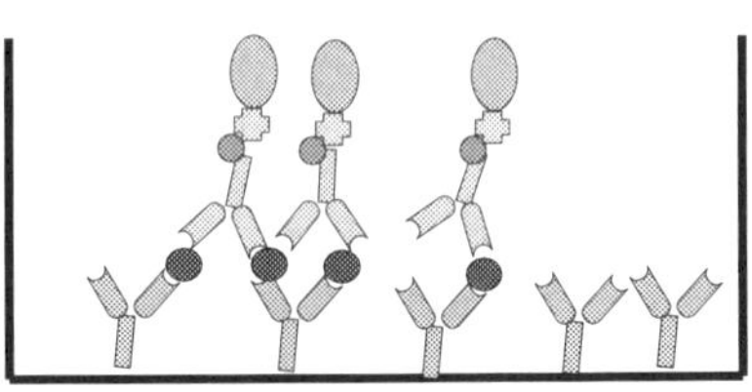

(c) In the biotinylated assay use is made of the strong specific interaction between biotin and streptavidin. The primary antibody is conjugated to biotin, and the avidin molecules are conjugated to the enzyme label.

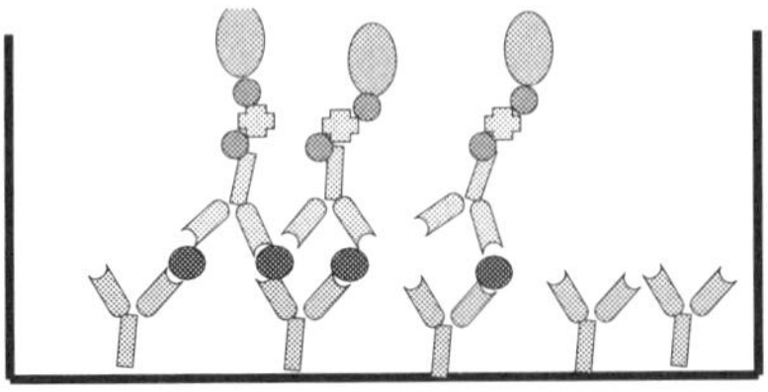

(d) An adaptation of the biotinylated assay in which the enzyme is conjugated with biotin and avidin forms a bridge between the enzyme and the antibody.

Fig. 7.2 Amplification of the ELISA.

coloured product had the advantage of being cheap and requiring fairly inexpensive optical density plate readers. Fluorescent or luminescent assays have the advantage of increased sensitivity, but the disadvantage of higher backgrounds, and expensive costs for the small laboratory or institution.

Standardisation

Reagents and procedures must be standardised in order to produce reliable results. Calibration and the appropriate quality controls ensure that the results can be compared between assay runs and between laboratories. Use of a single standard or reference preparation worldwide (such as that of the National Institute for Biological Standards and Control) allows universal comparison of results. However, at the time of this book going to press, none are available for food allergens, so appropriately stored in-house or secondary standards must be used. Assays must be validated for the types of samples to be processed. It is not sufficient to determine by experiment that an assay has a sensitivity of so many μg/ml for peanut in flour and then assume that it would have the same sensitivity for peanut contamination of chocolate. The sample matrix or composition may affect the assay, giving artificially high or low readout values. Spiking a pure preparation of known allergen with the analyte-free food will reveal what these effects are.

7.3.3 Comparison of allergen contents of different foods or food sources

ELISA with a standard curve provides information on quantities of specific allergens. Sometimes it is necessary to compare one extract with another to determine if the allergens present are the same (homologous) or different (heterologous). This is useful for determining if a particular treatment reduces or increases the allergenicity of a particular food source, for example whether hydrolysis has removed cow's milk allergens from infant formula. It is particularly useful for determining if a food causing an allergic reaction was contaminated or contains cross-reacting proteins with another known allergen, e.g. a food containing hazelnuts that caused a reaction in an individual with peanut allergy. Figure 7.3 shows a schematic representation of an ELISA inhibition to determine the similarity of two allergenic food sources. This assay may also be adapted to provide quantitative data where the percentage inhibition obtained with dilutions of a homologous antigen to the one bound to the plate is used as the standard curve. In this type of assay mouse monoclonal or rabbit antibodies are used rather than patient sera.

7.3.4 Extraction

The steps required are as follows

- Sampling
- Grinding and/or homogenising

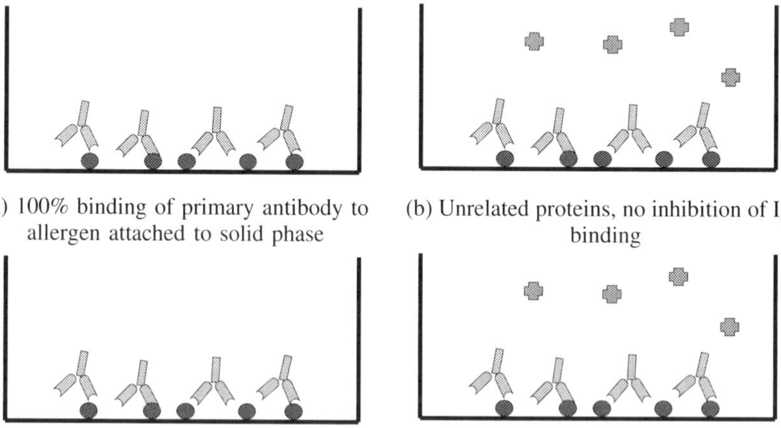

(a) 100% binding of primary antibody to allergen attached to solid phase

(b) Unrelated proteins, no inhibition of IgE binding

(c) Related proteins, inhibition of IgE binding

(d) Heterogeneous mixture with a shared protein

Fig. 7.3 Schematic representation of ELISA inhibition to determine the similarity of two allergenic food sources.

- Concentration and/or extraction into suitable buffer
- Removal of particulate matter.

The efficiency of the extraction procedure will vary for individual allergens and for different food matrixes. The expected recovery should be estimated by experiment in all circumstances.

Dry powders and cereals

Many food sources will be dry or semi-dry. In order to achieve adequate extraction the matrix must be broken down. This can usually be achieved by grinding, using a warring blender. Matrixes such as chocolate may be most easily treated by liquefying by heating and then extracted as a liquid using a warm buffer. Most allergens are water soluble and so can be extracted directly into the assay buffer. Common methods employ mixing (using a rotary shaker) the ground food with phosphate-buffered saline overnight at 4ºC. Once extracted, particulate material is removed by sedimentation with or without centrifugation and filtration where necessary.

Liquids

Liquids must of course be homogenised by mixing. It may be necessary to concentrate the sample. Common techniques involve dialysis to exchange buffers and/or remove low molecular weight contamination, followed by freeze-drying to concentrate. Proteins may also be concentrated by virtue of size using an Amicon filtration unit, or a Sephadex G25 column.

Oils
As the majority of allergens investigated are water soluble, the oil can simply be shaken overnight with an equal volume of aqueous assay buffer. The oil/aqueous layers are then separated by cold centrifugation and the aqueous layer decanted. Concentration can then be employed as above. Alternatively detergents may be used to extract the proteins, but these may interfere with the subsequent assays.

7.4 Processing and effects on allergenicity

7.4.1 Food processing
Some foods are encountered unprocessed, such as allergens from fruit and nuts. Foods such as cereals, egg and fish are, however, more commonly processed, and as mentioned earlier proteins vary in their stability. In some cases food processing practices have been developed in order to reduce allergenicity. Only the allergens from fresh fruits and vegetables are very unstable, such as apple allergen (Bjorksten *et al.* 1980), and are inactive after mild heating or even mashing. The majority show varying degrees of resistance to processing, the extreme being shrimp allergen that may remain active even in steam droplets.

7.4.2 Heat treatment and cooking
As a general rule heat decreases the allergenicity of proteins, and heat in the presence of moisture even more so, but this biological activity is rarely removed. Allergenicity of whole wheat flour or purified gluten is only reduced and not eliminated by heating up to 120°C for up to one hour (Varjonen *et al.* 1996, Sutton *et al.* 1982). Heating rice glutelin and globulin fractions also reduces IgE binding ability by 40-70% (Shibasaki *et al.* 1979), but the food remains allergenic. Peanut and nut allergens are resistant to heating and even roasting.

7.4.3 Hydrolysis
Hydrolysed casein and whey infant feeding formulas have been developed with the aim of reducing symptoms of milk allergy in infants. However, allergic reactions have occurred in some infants fed with these formulas, so tests have been developed to estimate residual activity. Hydrolysis is aimed at destroying the allergenic epitopes by cleaving the protein molecules into peptide fragments. Some are extensively hydrolysed and filtered, and it is becoming apparent that only these reduce the risk of atopic sensitisation.

7.5 Summary

At present only a limited range of detection kits are available commercially. As this is a rapidly expanding field the current manufacturers have not been listed.

Instead it is hoped that the reader will be equipped with an understanding of the techniques involved and what criteria should be specified for the assay. The following questions should be raised when considering the use of the various methods:

- What are the detection limits required?
- How sensitive is the assay under the conditions in which you wish to use it?
- How specific is the assay for the matrix in which you are trying to detect the allergen?
- What is the percentage recovery of your extraction procedure?

7.6 References

ASTWOOD J D, LEACH J N and FUCHS R L, 'Stability of food allergens to digestion in vitro', *Nat Biotechnol*, 1996 **14**(10) 1269–73.

BECKER W M, 'Characterization of Ara h 1 by two-dimensional electrophoresis immunoblot and recombinant techniques: new digestion experiments with peanuts imitating the gastrointestinal tract', *Int Arch Allergy Immunol*, 1997 **113**(1–3) 118–21.

BJORKSTEN F, HALMEPURO L, HANNUKSELA M and LAHTI A, 'Extraction and properties of apple allergens', *Allergy*, 1980 **35** 671–7.

BLANCO C, DIAZ-PERALES A, COLLADA C, SANCHEZ-MONGE R, ARAGONCILLO C, CASTILLO R, ORTEGA N, ALVAREZ M, CARRILLO T and SALCEDO G, 'Class I chitinases as potential panallergens involved in the latex-fruit syndrome', *J Allergy Clin Immunol*, 1999 **103**(3 Pt 1) 507–13.

BRADFORD M M, 'A rapid and sensitive method for quantitation of microgram quantities of protein utilizing the principle of protein-dye binding', *Anal Biochem*, 1976 **72** 248–254.

BUFE A, 'The biological function of allergens: relevant for the induction of allergic diseases?' *Int Arch Allergy Immunol*, 1998 **117**(4) 215–19. Review.

CABALLERO T and MARTIN-ESTEBAN M, 'Association between pollen hypersensitivity and edible vegetable allergy: a review', *J Investig Allergol Clin Immunol*, 1998 **8**(1) 6–16. Review.

HOURIHANE J O'B, KILBURN S A, NORDLEE J A, HEFLEE J A, TAYLOR S L and WARNER J O, 'An evaluation of the sensitivity of peanut allergic subjects to very low doses of peanut protein: a randomised, double blind, placebo-controlled food challenge study', *J Allergy Clin Immunol*, 1997 **100**(5) 596–600.

ROITT I M, BROSTOFF J and MALE D, *Immunology* 5th edn, Mosby, London, 1998.

ROSENBERG I M, 'Getting started with protein purification'. In Rosenberg I M (ed.), *Protein Analysis and Purification Benchtop Techniques*, pp. 99–109, Birkhauser, Boston, 1996.

SHIBASAKI M, SUZUKI S, NEMOTO H and KUROUME T, 'Allergenicity and lymphocyte-stimulating property of rice protein', *J Allergy Clin Immunol*,

1979 **64** 259–65.
SMITH P K, KROHN R I and HERMANSON G T, 'Measurement of protein using bicinchoninic acid', *Anal Biochem*, 1985 **150**(1) 76–85.
SUTTON R, HILL D J, BALDO B A and WRIGLEY C W, 'Immunoglobulin E antibodies to ingested cereal components: studies with sera from subjects with asthma and eczema', *Clin Exp Allergy*, 1982 **55** 256–67.
VARJONEN E, BJORKSTEN F and SAVOLAINEN J, 'Stability of cereal allergens', *Clin Exp Allergy*, 1996 **26** 436–43.

8

Handling food allergens in retail and manufacturing

J. Hignett, Nestlé UK Ltd, Croydon

8.1 Introduction

Food allergies can be uncomfortable, severe or potentially fatal to those who suffer them, depending on the nature of the reaction. The most common advice to sufferers is to avoid consumption of the trigger food in the diet. On the surface this seems a relatively simple and straightforward means of avoiding reactions. However, the fact that some individuals can react to minute amounts of the trigger food combined with the fact that the most common triggers of food allergies (milk, egg, wheat and nuts) can be widespread throughout a host of different foods means that avoiding allergens can be a time-consuming process.

All food manufacturers have an overriding legal responsibility to ensure that their products are safe and fit for the purpose intended. They must also comply with the relevant labelling legislation. The first step is to identify the key allergens. These are the allergens that are the most common causes of food allergies. Following this, a comprehensive evaluation of ingredients, storage, products and processes needs to be undertaken to understand in detail those products that contain these key allergens. Peanuts and nuts are considered as a special case in manufacturing and retail as they currently seem to be the major cause of anaphylaxis in the UK, a severe and potentially fatal food allergy. Peanuts themselves appear to be the most potent allergen and are the main cause of severe reactions.[1] They seem to initiate reactions in some peanut allergics at very low levels. Other nuts are also implicated in anaphylactic reactions and these include hazelnuts, almonds, brazils, cashews, pecans and walnuts. Additional controls at all levels are often introduced to ensure that the presence of even trace amounts of certain allergens in a product is communicated to sufferers.

Although both the retail and catering environments operate on a smaller scale than food manufacture, the principles involved in the handling of allergens are identical to that in large-scale food manufacture. The areas of concern when handling allergens and the actions that can be taken apply equally to large-scale manufacture as to smaller operations. However, one major difference is that food sold unpackaged, for instance in delicatessens and bakeries, is not sold with a detailed ingredients list. The handling of allergens in such cases mirrors that in the catering sector and is discussed in this chapter.

The main communication tool that the industry has is the ingredients list provided on the majority of products. It is important that ingredients lists are thorough, accurate and legible, and this is the duty of every manufacturer. However, this is not always straightforward and some of the issues surrounding this will be discussed in detail in this chapter.

8.2 Identification of allergens

The main challenge to the food industry is to provide accurate and up-to-date information at all times so that sufferers of allergies can select foods with confidence.

The first step in identifying a strategy for managing allergens in the food industry is to highlight the key allergens to be controlled. These do vary from country to country, but certain allergens are seen as important in most countries. Other key allergens may vary and can be added to the list to suit the specific country; for instance in France celery is viewed as an important allergen.

Suggested key allergens are as follows:

- Milk
- Wheat
- Eggs
- Soya
- Peanuts
- Tree nuts
- Shellfish
- Sesame.

Nut oils are an area worthy of comment. Research suggests that refined, bleached, deodorised nut oils do not initiate allergic reactions,[2] even in those who are anaphylactic to nuts. Unrefined, cold pressed or virgin nut oils are chosen specifically for their distinctive flavour. These oils are not refined and contain small amounts of protein from the original nut. The same research study showed that unrefined oils were able to cause reactions in individuals who have suffered previous anaphylaxis, but that generally the reactions were not as severe as those experienced when nuts were eaten. The small amount of protein that is present in unrefined oils is removed through the process of refining, bleaching and deodorising, rendering the oils suitable for allergy sufferers.

Once key allergens have been identified, all steps in the manufacturing process need to be cross-checked to ascertain whether there are any allergens present in the product, or indeed whether there is any chance of cross-contamination with any allergens during the manufacturing process.

8.3 Good Manufacturing Practice

Good Manufacturing Practice (GMP) in the food industry is the series of controls used during production that are aimed at ensuring that all products are consistently manufactured to a quality appropriate to their intended use. GMP aims to produce safe and wholesome food through well-controlled operations that avoid waste and any type of contamination. It should be applied throughout the whole production and supply chain and covers areas such as raw material sourcing, hygienic design of buildings and equipment, production processes, food handling, storage and transport conditions, safety procedures, cleaning procedures and personnel hygiene. The ability to demonstrate the principles and measures involved in GMP and the actions that are taken at a particular manufacturing site are essential to show that all reasonable steps are taken to prevent errors and indeed offences from potentially occurring. The manufacturer of a food product must comply with the relevant legal requirements, including product composition, labelling, safety and hygiene. GMP is an overall system for control and maintenance of quality. In its broadest sense it shows that quality is the responsibility not only of the factory, or group of factories, but also of suppliers, contract manufacturers and all business partners. The principles outlined in GMP have been developed for large-scale food industries, but they apply equally well to retail and catering environments, albeit on smaller scales.

Ideally, production facilities that handle ingredients containing key allergens will be specifically designed and built to enable complete segregation between products containing key allergens and those that are free from those allergens. A factory that produces food containing allergens should ideally have the following properties:

- Dedicated equipment
- Screened-off manufacturing/packing areas
- Dedicated workwear and washing facilities
- Cleaning regimes and pre-use inspections
- Segregated storage areas
- Air flow management/negative air pressure in nut areas.

However, in practice many manufacturing plants are generally used for the production of more than one product, and often one of the products contains a key allergen. Where dedicated equipment is not available for one particular product that contains key allergens, additional controls need to be introduced to control the presence of allergens and prevent contamination of other products with key allergens.

8.3.1 Allergens and GMP

The control of allergens in the food industry clearly falls under the remit of GMP, as key allergens in products should be labelled as any errors or omissions have the potential to cause serious safety problems for those who suffer allergic reactions. The areas of product composition, labelling and safety are particularly relevant to the control of allergens in the manufacturing process and these will be discussed in detail throughout this chapter. A number of criteria must be considered to state that a given product is free from a particular allergen.

To claim that a product is free from a particular allergen it must:

- not contain the allergen as an ingredient;
- not contain any rework or any other material that contains the allergen;
- not carry any risk of cross-contamination with the allergen through manufacture or packing on a plant where other products containing the allergen are processed.

8.3.2 Hazard Analysis and Critical Control Point (HACCP) studies

HACCP studies are detailed procedures which are undertaken to evaluate possible safety hazards, to eliminate them where possible or to find ways of keeping them under control, and are an important part of any Good Manufacturing Practice plan. These studies are not mandatory but are a useful tool in food manufacture to demonstrate diligent care during production. HACCP studies involve the identification of Critical Control Points (CCPs) in a manufacturing process using a systematic and standard approach to hazard analysis. CCPs are those specific parts of a manufacturing process where there is a risk of contamination of a product occurring and where a specific control needs to be introduced to minimise the risk. They range from critical points in the storage of raw materials to prevent cross-contamination, to the cleaning of a particular part of a plant following production, to the use or disposal of any waste that may be produced. CCPs can be related to microbial contamination, but in this case will be discussed in relation to contamination with allergens. Once CCPs have been identified, the risks need to be detailed and the procedures developed to minimise the risks of contamination. Training, reporting and documentation of the actions taken are also part of any HACCP study, to ensure consistency in quality control for every production run.

These studies are invaluable in the control of allergens in the manufacturing environment as they give a clear indication of the risk of allergens being present in a specific product, particularly through potential cross-contamination from or to other products. Although not mandatory, HACCP studies should be undertaken on each production line and are a critical part of any Good Manufacturing Practice procedures used in a manufacturing site.

8.3.3 HACCP in practice

Each step in a manufacturing procedure needs to be assessed from an HACCP angle. Consequently any manufacturing process will have many such assessments covering various aspects of the production chain. Critical control points that relate to the handling of common allergens cover all areas of manufacturing, including storage, production, packing, transport and employee safety.

In all cases, the HACCP study must include the identification of the CCP, providing a clear outline of the potential hazard, details of the control measures in place, identification of the person responsible for the control measures, the action required to ensure the controls are met and finally any corrective action should a problem occur. The HACCP study needs to be undertaken in a systematic and thorough manner. Each step in the process of manufacture, from receipt of ingredients to packing of the finished product, must be assessed. Table 8.1 shows one example of a systematic approach to an HACCP study.

Each HACCP study gives a detailed review of a specific step in the manufacturing process. An example of an HACCP study is given in Fig. 8.1, showing the handling of nuts in a production facility. The particular step involved is the disposal of packaging in which nuts are delivered to the factory. There is a risk that this packaging could be reused and could transfer traces of nuts to other ingredients.

This specific example shows the detail required for each critical control point in the HACCP study. A completed HACCP study provides an extremely thorough review of the entire manufacturing procedure and gives very detailed advice to the operators of the production line to control any risks that could arise. HACCP studies provide a very useful tool for quality control.

HACCP studies must be undertaken for each production line and must be recorded in detail. It is important that instructions and training are provided to all operators, outlining the steps that need to be taken to control any risks. In the case of completely new lines, new products being manufactured on an existing line, or major line modifications, the HACCP study must be repeated as even small changes to procedures can introduce new CCPs. The HACCP study must take into account the real-life characteristics of any line to provide accurate information and appropriate controls.

An essential part of all HACCP studies is a clear training programme to ensure that all staff who work in a particular area are fully aware of the

Table 8.1 Systematic approach to HACCP studies, related to allergen control

1 Process step details	What is the nature of the process involved?
2 Hazard Description	Is there a risk of contamination with allergens?
3 Control measures	What procedures will control this risk?
4 Modifications	Can changes control the risk – what are they?
5 Is it a Critical Control Point that needs documenting and controlling?	

HACCP – Nuts Critical Control Point No. 8 Disposal of packaging	
Hazard	Contamination of other lines if nut packing materials are used for other purposes, e.g. storing other raw materials
Controls	Prevent use of packaging for other purposes to ensure no cross-contamination to other ingredients
Person Responsible	Mixer operator
Action Required	Dispose of all nut packaging materials once used
Corrective Action	If nut packaging materials are observed in use for other purposes report to LINE MANAGER immediately

Fig. 8.1 Example HACCP study.

background to the HACCP studies and the actions that must be taken. Briefing about allergies should be included in all induction sessions and regular updates will ensure that staff continue to be well informed. It is particularly important that information is given to all those who work in an area that handles nuts and peanuts, and all staff understand the importance of the quality controls.

8.4 Control of allergens throughout the supply chain

8.4.1 Cross-contamination

Cross-contamination is the risk of small particles of one ingredient being transferred from a product where they are added to another product where that ingredient is not present. Although it is a term that sounds negative, from a food industry point of view it simply represents the risk of small amounts of certain ingredients being present in a product to which they were not initially added. This can occur when two or more slightly different products are manufactured or packed on the same line and have different ingredients, such as cereal products with different additions or different flavours of chocolate bars. Cross-contamination of ingredients or products can occur at the level of the raw material supplier (who may process many raw materials), during transport or storage of raw materials or, indeed, during manufacture or packing of the finished product.

In relation to allergens, cross-contamination is a real risk that must be controlled or acknowledged on the label. In most cases it is only minute amounts of an allergen that are transferred from one product to another. However, it is clear that very sensitive individuals can react to extremely small quantities of allergens, so cross-contamination of any nature must be handled properly.

HACCP studies, as detailed earlier, are used to identify any risks of cross-contamination, which can occur at any point within the supply chain. Where a risk exists there are two options, namely control of the risk or use of the appropriate labelling on the product. Peanuts and nuts are particularly common agents involved in cross-contamination and the statement 'May contain nut traces' can be seen on a number of products. The use of the 'may contain ...' statement is not a substitute for Good Manufacturing Practice and appropriate controls, and it should only be used where a real risk of cross-contamination exists. The most common product lines to carry 'may contain' statements are chocolate products, as chocolate is usually produced on a continuous process, and although cleaning of lines is undertaken between products a full cleansing is usually performed less frequently, as water and chocolate do not mix. Other areas that pose risks are those where dry ingredients are used and dust may be present in the atmosphere, as in breakfast cereal production.

Cross-contamination is not restricted to large-scale food manufacturing environments. The risk is equally problematic in bakery shops, small confectioners and out-of-home eating establishments. The use of tongs, scoops, dishes and trays is often common to a number of products in these areas. Think of purchasing a doughnut from a small bakery where the doughnut will be placed into a bag to take away using tongs. There is a risk that those tongs were last used to handle a Danish pastry that may have had nuts liberally sprinkled over the top or a cake with an egg-based icing. Even these minute quantities of allergens can pose a risk for very sensitive individuals. The control of allergens in these circumstances and the communication to the ultimate consumer is much more difficult.

All aspects of the supply chain must be evaluated for presence or risk of contamination with key allergens. This includes purchasing of raw or semi-finished materials, transport of these materials, storage within the production unit, production, packing and distribution. At each stage full HACCP evaluations of all equipment used, processes and risks need to be undertaken and documented to provide information on the suitability of the product for sufferers of different allergies. A full evaluation of a production line may involve many HACCP studies.

8.4.2 Purchasing raw materials

All raw ingredients such as flour, milk, nuts and fruit, and compound ingredients such as processed cereals, chocolate, biscuits or toffee must be purchased against a detailed specification. This must include the nature of the product, the ingredients included in a compound ingredient, and any risks of cross-contamination that may occur in the production or packing of the ingredient that is purchased. Supplier Quality Assurance is a system whereby suppliers are audited to ensure that they meet the high quality standards demanded by food manufacturers; it places the responsibility of meeting the standards set by the manufacturer clearly within the remit of the supplier. The initial audit procedure is a detailed analysis of the supplier and the operations that occur within their facilities. It is essential that it includes a detailed risk assessment relating to the presence of allergens and particularly the use of peanuts and nuts at the supplier's location. The presence of allergens in any raw material needs to be clearly acknowledged, even down to the level of carriers in flavours. Additionally, any real risk of cross-contamination from other materials that may be processed or packed in the same facility must be assessed in detail. HACCP studies can be used to identify any CCPs in the supplier's procedures. Where a real risk of cross-contamination with allergens exists, it must be highlighted on the specification to ensure that information is fed through and eventually highlighted on the label. This is particularly relevant for compound ingredients such as chocolates and cereals. The suggestion that cross-contamination exists must be ascertained following a detailed review and is not a substitute for Good Manufacturing Practice, nor should it be used to discharge any liability should a problem occur.

8.4.3 Distribution of raw and semi-finished materials

The distribution of raw materials from their site of packing or production to the factory where they are used also represents an additional risk area for cross-contamination. The transport of allergen-containing ingredients needs to be undertaken with care to ensure that traces of these are not transferred to other raw materials. They should be transported and stored in fully sealed containers to reduce any risk of spillage, and clearly labelled so as not to be confused with other ingredients. Any risks of cross-contamination need to be identified and noted so they can be marked on the label of the finished product. Colour coding of packaging is a useful way of segregating and identifying allergen-containing ingredients. A standard colour can be used for the packaging and containers in which the ingredient is stored and for the equipment associated with the production of products containing that ingredient (such as trays, moulds, dishes and brushes). A colour coding system must be robust to ensure that confusion does not arise. A standard colour should be chosen for use through the supply chain and this should be applied and adhered to rigidly. Induction sessions to new employees must include detailed instructions on any colour coding procedure.

8.4.4 Storage of raw materials

Storage represents another risk area for cross-contamination. High-risk ingredients (from an allergy perspective) need to be stored with caution to prevent any cross-contamination occurring from spillages, poor labelling or even absent labelling. Ideally, high-risk ingredients, such as nuts and nut-containing ingredients in particular, should be stored in locked storage areas and be accessible only by authorised personnel when required for use in the production facility.

8.4.5 Scheduling

Production schedules give detailed information on the precise nature of a product to be manufactured on a given plant at a particular time, and the programme of products to be made over a given period. In many cases production lines are used for the manufacture of a number of products, including different flavours of one product or completely different products altogether. Schedules can be planned to reduce the risk of transfer of allergens from one product to another. Plants that are used to produce more than one product may be cleaned down after production of each product has been completed, with a full and thorough cleansing undertaken at regular intervals. Products that have a higher risk of microbial contamination, such as those containing fresh milk and raw meat, will be cleaned down thoroughly much more frequently. Effective management of production schedules will ensure that products containing allergens will be manufactured at the end of the programme before a full cleandown. Additionally, allergen-free products should be manufactured immediately after a full cleandown to minimise any risk of transfer of allergens to products that are manufactured later in the schedule.

8.4.6 Manufacturing

The control of allergens during manufacture and packing is a critical area. HACCP must be used at each stage of the manufacturing process itself to ascertain key areas of risk. Additionally, all equipment associated with plants that manufacture products containing allergens must be controlled to ensure there is no risk of transfer of allergens on brushes, spatulas and other items. A colour coding system is the best way of easily identifying equipment associated with a particular plant. One colour can be used for all items associated with that product such as trays, moulds, brushes and rework containers.

Manufacturing plants are often complex with many different parts that are capable of harbouring allergens. HACCP studies will identify the key areas of risk and appropriate controls can be introduced, as discussed earlier. It is essential that all staff are briefed on the importance of the controls introduced and take responsibility themselves for the quality control of the products they are manufacturing. Routine sensory evaluation undertaken on newly manufactured products is a useful tool to monitor quality control. Trained panels who are expert assessors on a particular product are an excellent resource to confirm that a given production run meets the specification for that product. They are also capable of identifying problems with a product and may pick up a problem associated with cross-contamination of allergens.

8.4.7 Rework

Rework is a term given to slightly defective or excess product or ingredients that are newly processed but not suitable for packaging immediately into finished product. It is first checked to ensure it is of a very high quality, and can be reworked and added back to products. Lower quality waste is not added back to products but disposed of in an appropriate manner. Controls must be in place to ensure there is no cross-contamination of allergens when using rework. The simplest rule when handling rework is to put 'like into like' to prevent any risks. Additionally, rework must be clearly labelled for further internal use within the factory and controls must be in place to ensure it is used correctly. Rework is particularly an issue with regard to nut allergy as trace amounts of an allergen can easily be transferred.

8.4.8 Air movement

Handling dry allergens such as powders, nuts and dusts creates additional risks associated with the movement of air that may carry particles of allergens. Since a small number of sufferers react to extremely small amounts of allergens, care needs to be taken where there are excessive amounts of dust to ensure other products are not 'contaminated' with this dust. Air conditioning needs to be installed to prevent air containing dust particles being transferred to a separate part of the production facility and allergens being transferred with it.

In extremely dusty environments, such as nut roasting plants, additional care needs to be taken to prevent allergens being transferred on clothing from one part of the factory to another. On entering the high-risk area employees and visitors should be required to wear special protective clothing such as full body suits and hair cover. On leaving the specific area the protective clothing should be removed to prevent any transfer of allergens to another area of the factory.

8.4.9 Employees and visitors at manufacturing locations

In addition to the risks associated with products containing allergens, risks also arise to personnel who themselves are allergic to certain ingredients and who are employed at, or visit, specific manufacturing sites where these ingredients are used. All employees should complete a pre-employment questionnaire and medical to ascertain whether any suffer food allergies and particularly anaphylactic reactions. Those that do suffer should not be expected to work in areas where allergens to which they react are processed.

All visitors to a site where key allergens are used should be informed about the nature of products manufactured at that site and informed that if they are allergic to any ingredients used they are advised not to visit the manufacturing plant. Once again this is particularly relevant to the use of nuts and peanuts as reactions can be so severe. A suggested outline for a notice at the reception desk is provided in Fig. 8.2.

8.4.10 Canteen and restaurant facilities

The control of allergens extends from the production line itself to all areas of food provision within the manufacturing site. This includes canteen and

PLEASE READ BEFORE SIGNING IN

Welcome to our factory. This site uses peanuts and hazelnuts in some products. Peanut or hazelnut dust is present in the air within certain areas of the factory. If you suffer allergic reactions to either peanuts or hazelnuts we recommend that you do not enter these areas of the factory at all. Please speak to your contact at this site for information regarding areas where peanuts and hazelnuts are used. If you are in any doubt at all regarding your susceptibility to peanuts or hazelnuts we advise you to avoid these areas.

Fig. 8.2 Typical factory advice sheet.

restaurant facilities as well as snack and coffee bars in factories and retail environments. All the controls outlined here for food manufacture apply equally well to food provision in a catering environment. Allergens should only be present in products where a sufferer would expect to find them, and information should be available for allergy sufferers to consult to assess whether a certain dish is suitable for their specific diet. If in doubt the allergy sufferer should be advised to avoid the dish and choose another option. The handling of allergens in the catering trade is discussed in more detail later in the chapter.

8.4.11 Confirmation of presence of allergens

Once all the above steps have taken place, food manufacturers are able to make a judgement based on all the evidence obtained as to whether a product contains or is free from a particular allergen. Information should be provided to allergy sufferers to enable them to select suitable foods for their diet. The provision of information to consumers on packaged food and food sold loose is discussed later in the chapter. In addition, a number of tests are available that can be used to analyse products for the presence of a given allergen. Generally a radio-immunoassay technique is used which checks samples of a product for specific proteins that have been previously identified as allergens. These tests can be useful, but in some instances results do need to be interpreted with care. Any analysis is only as accurate as the samples that are taken. The sampling of liquid or fluid foods gives a relatively reliable sample, as the food can be further blended to give an even distribution of all ingredients. The sampling of foods such as breakfast cereals, chocolate bars and other more complex foods poses a number of difficulties. A number of samples could be taken randomly from the food according to good practice, but there is a chance that the one small piece of allergen, be it a flake of nut or a grain of milk powder, could be missed. The results achieved would give a false negative, suggesting that a product is free from a particular allergen, as a random sampling technique did not actually pick up the small amount of allergen present. Such tests should not be used to give definitive information about the presence or absence of allergens in a product.

Their use complements the results obtained from a full HACCP study that should be undertaken on each product.

8.5 Other initiatives

Most food products sold through retail channels are packaged in such a way that ingredients lists on products provide an easy way for allergy sufferers to check the suitability of that product for their particular diet. Where products are sold without packaging or the packaging is removed before being presented to the consumer, the communication of the suitability of that product for allergy sufferers becomes more difficult. Both the catering trade and some areas of retail are areas where the communication of the suitability of products for allergy sufferers is extremely difficult, as foods are sold without labels showing the detail of the ingredients they contain. Allergy sufferers must take it upon themselves to check the suitability of any foods for their particular diet, and if in doubt at all about a particular product or dish they should avoid it.

The Ministry of Agriculture, Fisheries and Food has prepared a list of guidelines for catering establishments to raise awareness of the issue of food allergies and to help caterers provide information for sufferers. This is equally applicable to small retail environments. An extract from the recommendations is provided in Fig. 8.3.

8.6 Key aspects of legislation from a manufacturing view

Food legislation plays an important role in the development of a policy for the handling of allergens in food manufacturing. Manufacturers have a responsibility to provide safe food for consumers, and this includes safety from an allergy sufferer's view. The obvious legislation is that which directly relates to food, such as Food Labelling Regulations. However, in addition other areas of the law need to be considered and these include relevant consumer protection legislation and requirements arising from the European Product Liability and Product Safety Directives. Manufacturers need to consider the extent to which their position and that of their products will be influenced by a number of potential circumstances.

A significant number of people have unfortunately died or have been seriously ill as a direct result of an allergic reaction following the ingestion of foods which, unknown to them, contained small amounts of allergens to which they had an anaphylactic reaction. These cases have received widespread publicity. Responsible food manufacturers know that a number of foods and ingredients can give rise to rapid, life-threatening reactions in a small number of allergic individuals. The adverse publicity that might be received following an incident could be extremely damaging to the reputation of the product concerned and, indeed, the company's standing.

Advice to catering establishments

In case a customer asks you about the ingredients of a meal, you should aim to make sure that there is always someone on duty who knows or can find out the ingredients of all the foods you provide. If you are not sure whether there is a trace of a life-threatening ingredient in a meal then say so – never guess. If foods contain nuts, make sure this is reflected in the name or the menu description, for example, carrot and nut salad.

Foods to watch out for

Many establishments often use nuts and seeds to decorate cakes, ice creams, speciality breads or savoury dishes. Other less obvious sources of nuts and seeds are:

- marzipan which is made from almonds
- hummus which contains sesame seeds and halva which is made from sesame seeds
- sauces such as satay sauce which is made from peanuts
- products such as Waldorf Salad, salad dressings and flavourings

A customer suffering from severe food allergy will usually know about the foods they must avoid

What your staff can do to help customers

If you are asked by a customer you must

- Tell them what is in your food – exactly
- If you don't know don't guess – find out!

Remember!

Even tiny traces of these foods can kill.

- Think before using nut and seed oils, salad dressings and seafood sauces
- Don't let nuts, seeds and shellfish touch food that shouldn't have those ingredients
- Clean your hands, work surfaces and utensils after handling nuts, shellfish and seeds
- Think before cooking with oils that have been used to cook other foods.

Fig. 8.3 Extract from MAFF guidelines: *Be Allergy Aware – Advice for Catering Establishments*.[3]

Manufacturers need to consider their legal obligations to inform purchasers of a product of the known or adventitious presence of allergens, even where this is not a specific requirement of current food labelling legislation. There are many additional points to consider from a manufacturer's view and these include some of the following. Is it appropriate that all relevant information should be given in the ingredients list or is an alternative location preferable? Where the presence of allergens is highlighted on labels the prominence of that message needs to be ascertained. There are not necessarily clear responses to these points available in law. It is clear that where an ingredient is knowingly added to a food, the Food Labelling Directive (79/112) requires that its presence should be declared on the ingredients list. However, there are exceptions to this Directive which are discussed in detail below.

8.6.1 Food safety legislation

Under the Food Safety Act 1990 all food manufacturers, caterers and retailers are required to ensure that the food they supply is safe (for the majority of people and when consumed in normal quantities) and is of the nature, substance and quality demanded. The General Product Safety Regulations 1994 (GPSR) will apply to food where there is no specific provision under the Food Safety Act (FSA) or any regulations made thereunder. As a result, information may need to be provided to consumers on any risks that a product might present regarding a number of factors, such as the effects of such a risk on a vulnerable group, for example allergic individuals.

A failure to comply with these requirements because of the unnotified, inadvertent presence of an allergen in a product through manufacture or cross-contamination, could give rise to a criminal offence being committed, even though no intention existed. There is, however, a due diligence defence available to manufacturers in the event of proceedings under both the FSA and the GPSR which would require the manufacturer to prove that he had taken all reasonable precautions and exercised all due diligence to prevent inclusion of an allergenic material. Manufacturers can reduce the risk of prosecution and contribute substantially to the establishment of a due diligence defence by implementing Good Manufacturing Practice and documenting all procedures taken as evidence of GMP processes, training and results, as detailed earlier.

8.7 Labelling and promotion

The majority of manufactured and packaged food products have to carry a full list of the ingredients they contain by law. The list shows the ingredients in descending order of weight in the finished product. There are currently no provisions made under either UK or EU food legislation which require potential allergens to be labelled. Whilst there is a general requirement that all ingredients added to a food must be declared on the ingredients list, in accordance with the

Food Labelling Regulations 1996, there are certain exceptions to this general rule. These relate to compound ingredients (an ingredient with a common name composed of multiple ingredients) which constitute less than 25% of the finished product, or to cases where the ingredient itself does not require an ingredients list if it were to be sold alone as a prepacked food (see exceptions below). Other exceptions to the Food Labelling Regulations include generic terms (e.g. fish can be used for any species of fish); 'carry-over' ingredients such as additives which do not have any technological function in the end product; additives used as processing aids; solvents/media for additives or flavourings; and those products which do not require ingredients lists at all such as food sold through catering outlets.

There are certain exceptions to the law. These include honey, condensed milk, dried milk products, coffee and coffee products, spreadable fats and chocolate. Each of these has its own regulations and needs to be considered individually.

8.7.1 The '25% rule'

This rule is contained in European Food Labelling Legislation. It states that compound ingredients (i.e. those that themselves contain a number of ingredients, e.g. toffee, biscuits, chocolate chips) that comprise less than 25% of the finished weight of the product need only be declared as the compound ingredient and not as the constituent ingredients that make up the product.

Manufacturers recognise the importance of providing information on the ingredients list to help sufferers of food allergies to select a suitable diet with confidence. To do this the list must accurately reflect the ingredients in the product, including those allergens that are present in minute amounts. Consequently, the majority of manufacturers voluntarily ignore the exceptions to the law and voluntarily label the presence of all allergens on the ingredients list. This includes carriers of ingredients, constituents of compound ingredients, and ingredients that may be present through cross-contamination that are on the list of key allergens.

8.7.2 A European view

Within the European Union various Member States are beginning to address the issue of labelling of allergens with various degrees of official recommendations. In France, the authorities have published a detailed review of the situation and recommended a number of ways in which industry, collectively, can significantly improve the information given to consumers. A restricted list of allergens is covered but the principal focus is on peanuts and similar derivatives, coupled with clearly defined changes to the legal framework for labelling. In Sweden, labelling legislation requires ingredients known to cause intolerance to be stated in the list of ingredients. Examples quoted include eggs, milk, gluten-containing grains, and legumes such as soyabeans, peas and peanuts. The

Swedish legislation also requires that when such ingredients are themselves present in compound foodstuffs, then any exemptions from ingredient declaration do not apply. In the UK, allergens do not have to be specifically labelled, though most manufacturers voluntarily provide information on the presence of peanuts and nuts where they are present.

8.7.3 International trade

The progressive development of international trade is leading to an increasing number of products sold with multilingual labels produced in one, or perhaps two 'European' factories for sale in several countries simultaneously. This situation is no longer confined to large multinational manufacturers but also applies increasingly to major retailers who, in some cases, are now selling products with European labels. This creates a number of problems from a labelling stance. Firstly, where two factories produce the same product there may be a difference in the other products manufactured at both sites and consequently a potential difference in the allergens that could be transferred by cross-contamination. It is essential that the 'worst case' scenario is alluded to on the label. For instance, where manufacture is split between two sites and one line also produces nut-containing products where there is a real risk of cross-contamination, this should be alluded to on the labels of both. This ensures consistency of labelling and removes any risk of confusion or any inadvertent consumption of a product that may initiate an anaphylactic reaction. Secondly, it is known that awareness of, and sensitivities to, different allergens do vary throughout Europe. The voluntary labelling of particular allergens is specific to some countries, whilst for others this additional labelling is not deemed as important in their country. Potentially, the presence of an allergen could be mentioned in one language but not in another, and this is an issue that individual companies need to address. However, by far the majority of products sold in a given country provide the ingredients information in one language. Nevertheless, this raises a further issue that companies need to address. Imported products need to conform to the legislation and the voluntary labelling actions taken in the receiving country that sells the product, which may differ from that in the manufacturing country. This will ensure that consumers have information they need to select products for their diet and can choose products from a given company with confidence.

8.7.4 'May contain' statements

The statement 'may contain xxx traces' is used to show where there may be small amounts of the allergen present in the product, most likely as a result of cross-contamination. Currently it is most commonly used for peanuts and nuts. The statement must only be used where there is a real risk of cross-contamination and not as a catch-all to remove any liability. GMP and HACCP studies will identify real areas of risk and the need to use such a statement. Where it is used it needs to be clearly legible and in a place where consumers

Ingredients

Wheat flour, sugar, hydrogenated vegetable oils, cocoa powder, modified starch, dried egg, dried skimmed milk powder, raising agents (E500, E450(a)), salt, flavouring, water, chocolate (contains lecithin & vanilla), acetic acid.

May contain nut traces

Fig. 8.4 'May contain' statement.

would expect to find it. It has become common practice within the UK to place this statement at the end of the ingredients list and, where possible, in a typeface slightly larger than that used for the ingredients list and similar to that used for the word 'Ingredients'. Figure 8.4 shows an example.

The use of 'may contain' advisory labelling in respect of the potential, adventitious presence of a food allergen should be a last resort. Such labelling should never be used as a general insurance and a substitute for Good Manufacturing Practice.

8.7.5 Brand extensions

Many brand names are now used across a wide variety of products; for example, a chocolate bar brand may be used for a dessert, ice cream, drink, chocolate spread, Easter egg, and various shapes and sizes of chocolate bars. It is possible that individuals with a specific food allergy and for whom the original chocolate bar is acceptable may assume that the other products sold with the same brand name are also suitable for their diet. However, in most cases different products will contain different ingredients, be manufactured on different production lines, in different factories, using different technologies and may well contain different allergens from other products under the same brand. It must be stressed that each product needs to be assessed on its own merits by the consumer by checking the ingredients list on the label. The onus is certainly on the consumer to check the suitability of each product for their particular diet.

8.7.6 Promotional activities

The control of allergens in manufactured products extends beyond production and labelling to all promotional practices linked with that product. Those that need particular attention are those that relate to sampling of the product. Product sampling can follow a variety of routes, but the most common include:

- Wet sampling – the product is served from a central location in a ready to eat or drink format, for immediate consumption.
- Dry sampling – a product that needs preparation is distributed from a central location in a format that needs further preparation.

- Door drop – free samples of products are distributed via the postal system for trial at home.

It is essential that those who are sampling products are fully briefed as to the allergenic potential of that product. Wet sampling of products, or the sampling of products intended to be consumed immediately, needs to be undertaken with great care, as consumers receive the product without any packaging. Information must be available to advise consumers of the ingredients in the product, and notices outlining any key allergens assist sufferers of allergies in selecting whether to sample that product. These procedures apply to dry sampling also, but in these cases the product is often distributed in its outer packaging with a detailed ingredients list. Sampling to children can pose additional difficulties and should only be undertaken with parental consent for the child to take the product. This is particularly relevant with nut and peanut allergies, as the reactions can be severe to extremely small quantities of the allergen.

Door drop sampling does provide an efficient way of inviting a large number of people to try a product. It, too, has difficulties. In households where someone suffers anaphylactic shock to a particular ingredient, the entire household very often follows the same principles and becomes an egg-free, milk-free or nut-free zone, for example. In such households, great care is taken to select foods that are free from the particular allergen to minimise any risk of anaphylaxis occurring. This is particularly true in households with young children who are unable to read labels and unable to be responsible for the foods they choose. It is also the case in many households where there is a sufferer of peanut or nut allergy, as these foods can be more easily taken out of the diet of the whole family than foods such as milk, eggs or wheat. Delivering free samples of foods containing the allergen through the letterbox removes the choice to select suitable foods from the family. A young child could see the food product on the doorstep and consume some without parental knowledge. Consequently, it is recommended that door drop sampling is undertaken with great care and is avoided entirely for products that contain nuts and peanuts. There are alternative options, including distributing a coupon for the product enabling sufferers of allergies to choose whether to sample that product, or a reply-paid card which is returned if the household would like to request a sample of a particular product to be delivered. The latter two mechanisms put the choice directly in the hands of the householders and remove any risk of inadvertent consumption of a product by young children.

8.8 Additional communication initiatives

The ingredients list on the label of a product is the most accurate way of assessing the suitability of a product for a sufferer of allergies. However, reading labels is a laborious and time-consuming process and makes shopping a lengthy ordeal. Most companies and retailers now produce lists of products free from key allergens which make food selection much quicker and easier. The lists are available from the companies directly and are often on the Internet. Once again peanut and nut allergies are often handled as a special case, as they are the most common food causes of anaphylaxis. 'Free-from' lists are updated every six months to reflect any changes that may have occurred. Users of lists are also advised to check ingredients' lists, particularly where a 'new recipe' or 'new improved' flash indicates a recipe change. In the case of anaphylactic reactions information must always be accurate and up to date. Peanut and nut-free lists are often controlled closely and carry a 'Use by' date after which that list is invalid and recipients are asked to contact the company for an update. During the 'shelf life' of the list it is recommended that the names and addresses of all recipients are held. Should any changes occur to that list whilst it is 'live', all recipients can be contacted to advise them of the changes and will be issued with a new list and asked to discard the old one. The distribution of the list to third parties such as dietitians and doctors is not supported, as this removes control of the list from the company. If a change occurred to a list, the company would rely on the health professional remembering which patients had received the list from them to pass on the update. These detailed procedures ensure that the company has tight control over this list at all times and can do everything they can to help sufferers of allergies to select suitable foods with confidence.

8.8.1 Food intolerance databanks

Many countries throughout Europe have food intolerance databanks managed by a central group, with information provided by companies. They collate information from various food manufacturers and produce comprehensive lists of products free from the key allergens. In many cases the booklets they produce (milk free, egg free, etc.) are available to health professionals, especially dietitians, who are then able to work with sufferers of food allergy to help them select suitable foods and also meet their nutrition requirements. The lists provide useful compilations of products suitable for particular diets, but are not without their pitfalls. Often they are updated only on an annual basis and risk becoming out of date even whilst they are still being issued. Additionally, they are not suitable for information on nuts and peanuts, as such information can quickly become outdated and is more dangerous than useful, for the reasons outlined above.

8.9 Summary

The management of food allergens in the food industry is a complex and time-consuming process, but one that is essential. The main aim of an allergen-handling process is to be able to provide accurate information to sufferers to enable them to choose a suitable diet with confidence. The detailed knowledge of the allergens used in a particular product, on a specific production line and in the factory site is the first step in assisting sufferers.

The key steps in managing food in food manufacture are, firstly, to understand the constituents of all raw materials in detail, secondly to check all procedures used during the manufacture of the product for any risks of cross-contamination of allergens, and finally to provide accurate information to consumers of the product regarding the allergens the product contains. All steps need to be undertaken thoroughly to ensure that even trace amounts of allergens are detected. The processes involved in Good Manufacturing Practice and HACCP studies assist in this process.

It is well known that sufferers of anaphylactic shock can react to extremely small quantities of allergens and it is for these people in particular that information provided about the suitability of the product for particular diets must be accurate and up to date. The labelling of packaged food provides the best communication tool, and the onus must be on the sufferer to check the labels of products to ensure suitability for their diet. Manufacturers must take responsibility to ensure that the labelling accurately reflects the ingredients in the product and any allergens that may be present through cross-contamination during the manufacturing process.

The communication of the presence of allergens in food sold loose without ingredients lists and food sold through catering outlets will continue to be a critical area. Continually raising the awareness of allergen control in these areas is a key task to ensure that those who suffer food allergies are able to select foods and meals with confidence.

The control of allergens in future will continue to be an important aspect of quality control for all aspects of food manufacturing, including large-scale manufacture, smaller-scale operations and catering processes.

8.10 Sources of further information and advice

Anaphylaxis Campaign
PO Box 149, Fleet, Hampshire GU13 0FA

British Allergy Foundation
Deepdene House, 30 Bellegrove Road, Welling, Kent DA16 3PY

British Dietetic Association
7th Floor, Elizabeth House, 22 Suffolk Street, Queensway, Birmingham B1 1LS

British Nutrition Foundation
High Holborn House, 52–54 High Holborn, London WC1V 5RQ

Food and Drink Federation
Federation House, 6 Catherine Street, London WC2B 5JJ

Leatherhead Food Research Association
Randalls Road, Leatherhead, Surrey KT22 7RY

Further reading

Food and Drink Federation, *Food Allergens Advice Notes*, FDF, London, 1998.

Institute of Food Science and Technology, *Food and Drink Good Manufacturing Practice – A Guide to its Responsible Management*, IFST, London, 1998.

Jardine N J (ed.), *Food Allergy and Other Adverse Reactions to Food*, International Life Sciences, 1994.

Lessof M (ed.), *Food Allergy: Issues for the Food Industry*, Leatherhead Food Research Association, 1997.

Nestec Ltd, *Nestlé Good Manufacturing Practice*, Nestlé, Switzerland, 1996.

Nestlé UK Ltd, 'Peanut and nut allergy', *Professional Care of Mother and Child*, December 1998.

8.11 References

1 COMMITTEE ON TOXITY OF CHEMICALS IN FOOD, CONSUMER PRODUCTS AND THE ENVIRONMENT, *Peanut Allergy*, Department of Health, London, 1998.

2 HOURIHANE J *et al.*, 'Randomised double blind crossover challenge study of allergenicity of peanut oils in subjects allergic to peanuts', *British Medical Journal*, 1997 **314** 1084–8.

3 MAFF, *Be Allergy Aware – Advice for Catering Establishments*, London, 1997.

9

Support organisations for individuals with food intolerance

D. Reading, The Anaphylaxis Campaign, Fleet

9.1 Introduction

In October 1993, my 17-year-old daughter Sarah died of an overwhelming allergic reaction after going into a restaurant and eating a slice of lemon meringue pie containing crushed peanuts. Sarah had thought she was only mildly allergic to peanuts and had no idea that an allergic reaction could kill. National newspaper and television reports referred to 'a very rare allergy to peanuts'. But its rarity was challenged by letters which subsequently appeared in a few of the newspapers: letters written by the parents of children with nut allergy. What was significant was that these parents had received little medical guidance about their children's allergy; they were coping alone. It also became clear there was a similar lack of knowledge and information within the food industry, and manufacturers and retailers had little or no idea that a major issue was about to break.

It was to fill these gaps in knowledge that I and a small group of parents launched the Anaphylaxis Campaign early in 1994. In a perfect world, there would be no need for organisations like ours. The government and medical bodies would take all necessary measures to ensure that anyone in the community with a particular medical condition would possess adequate information to understand it, manage it and, where appropriate, treat it. But the world is a complex place governed, to a large extent, by limitations of finance and resources. Moreover, those working in the scientific and medical community admit they do not have all the answers. We must accept that outside bodies have an essential role.

Support organisations working in the field of food allergy and intolerance have many functions, but a major priority must be the provision of information and guidance to people affected. This may be written information, in the form of leaflets, brochures, booklets and newsletters, or it may be verbal. The Anaphylaxis Campaign and the British Allergy Foundation both operate telephone helplines. These organisations regularly encounter deep anxiety among families affected by allergies, particularly where those allergies may be life-threatening, and testify to the fact that ignorance of the facts is usually the root cause of this distress.

Outside the family, there is a further demand for information. Children go to school, adults go to work, almost all of us visit shops and restaurants. We all have to eat. And so support groups face the interesting challenge of ensuring that knowledge and awareness are spread through all relevant sections of society, including schools, colleges, playgroups and, of course, food companies. Again, information leaflets and booklets are an important part of this education campaign, but often the most effective way of conveying relevant messages is through the media. This process may be proactive or it may be reactive. Many support organisations see the value in issuing press releases whenever they wish to promote a particular point, but it is more common for them to be called upon for a reaction to a current story. The difficulties in dealing with the media are obvious. Journalists are only human and may have only a five-minute telephone call in which they can get to grips with complex issues. There is always great potential for a misunderstanding; there is also the pressure from the newsdesk to stretch a point so that the original facts become distorted. In this way we get headlines such as '60,000 kids at risk from peanuts' – a serious overstatement in terms of the actual number whose lives are threatened. Despite the dangers, we must accept that the media is an essential tool at our disposal.

Educating the food industry is easier, at least in terms of getting the undistorted facts across. In my experience, there has been a gratifying willingness in most sections of the industry to face up to the problems of people with allergies. The deaths of five people during late 1993, and others since then, alerted the food industry to a problem which could not be side-stepped. I have no doubt that self-preservation among food companies has been a strong motivating factor, but I am equally convinced that there is genuine concern for those at risk. After all, even the most cynical food company director may have a family and identify closely with those parents who fear for their children's lives. Support organisations in the UK have had strong and fruitful discussions with food companies and their umbrella organisations since early 1994 and we have noticed a remarkable improvement in awareness and labelling. Things are far from perfect, but we're getting there.

Lobbying government departments and other statutory authorities is another important role of support organisations and, again, there has not been the uphill struggle that might have been expected. As I will describe later, support group representatives have held frequent discussions with government departments and there has been encouraging progress, particularly in the areas of food labelling, schools and research.

We exist, too, to raise awareness of allergy issues within the medical profession. There have been occasional alarming tales from patients reflecting a deep-rooted ignorance of allergies, particularly in doctors' surgeries. Many doctors may resent the intrusion of amateurs, but must surely accept that knowledge and understanding have plenty of room for growth. To add to the present pool of medical knowledge, many support groups have within their membership a large supply of information. By keeping the issues boiling, support organisations inevitably point to the gaps in understanding and identify where research is needed. Indeed, many charities working in medical fields have their own research funding programmes.

Having provided an overview of the function of support organisations, I will seek during this chapter to describe what they have achieved and what they expect of the food industry.

9.2 Current support organisations

Because allergy may affect up to 30% of the population, it is inevitable that the support organisations representing them are numerous. In the following section, I will introduce some of the leading groups working nationally.

9.2.1 The British Allergy Foundation

The British Allergy Foundation has a broad sphere of interest, encompassing all types of allergy. BAF was formed as a registered charity in 1991 by a group of leading medical specialists who were all determined to improve the awareness, prevention and treatment of allergy. The charity is managed by a board of trustees which deals with all the business aspects of the organisation. All decisions on medical and scientific matters in which the foundation is involved are made by a team of medical advisers. These are among the leading allergists in the country and most are members of the British Society for Allergy and Clinical Immunology.

The British Allergy Foundation is based at Deepdene House, 30 Bellegrove Road, Welling, Kent DA16 3PY and provides those affected by allergies with information and advice, including details of National Health Service allergy clinics. Leaflets, fact sheets and regular newsletters contain practical and informative articles. BAF also has a helpline (020 8303 8583).

BAF's membership stood at around 6000 by early 2000. Members are charged £10 a year. Regional contacts provide support for people with allergies and there are occasional meetings at which speakers keep people up to date and offer advice. Occasional research grants are made by the foundation in its attempts to discover the causes and hopefully new treatments. The foundation works closely with general practitioners and the National Health Service.

9.2.2 The Coeliac Society

The Coeliac Society was founded in 1968 as a national support group for people with coeliac disease and dermatitis herpetiformis. Coeliac disease is a lifelong inflammatory condition of the intestinal tract which affects the small intestine in genetically susceptible individuals. This is caused by gluten, a protein in wheat, and similar proteins in rye, barley and oats.

Through its national office in High Wycombe, the charity provides advice and support to people who have been, or are in the process of being, diagnosed. Queries are handled by an expert team, many of whom are themselves coeliacs. Where appropriate, members' concerns are referred to dietetic advisers. The society says that it does not interfere with advice given by a patient's own doctor but is able to give general support and guidance. Additional help is provided by local support groups led by volunteer organisers.

As well as information leaflets, the society publishes an annual booklet listing 10 000 gluten-free manufactured foods with updates provided throughout the year.

The society, whose address is PO Box 220, High Wycombe, Bucks HP11 2HY, maintains contact with government departments involved in healthcare and in food labelling legislation, and with European and international organisations supporting coeliacs. It funds research and development projects.

9.2.3 The Anaphylaxis Campaign

The Anaphylaxis Campaign, of which I am director, was set up early in 1994 following five well-publicised deaths caused by allergic reactions to peanuts or tree nuts. Those who died included my teenage daughter Sarah, whose death was particularly shocking because her previous allergic reactions to nuts had been mild. As a journalist, I had some expertise in gathering information and there were indications early on that, far from being rare, nut allergy was really quite common. Supported by my MP, Cranley Onslow, I set in motion the beginnings of an awareness campaign. However, I was not alone. Following the intense national publicity, several parents of children with nut allergy came forward and we formed the core group of the Anaphylaxis Campaign. As knowledge of the group spread, we found we were overwhelmed with letters from families similarly affected: 60–70 per day in the first few weeks. By early 2000, membership stood at around 5500. Members pay £5 a year.

The Anaphylaxis Campaign, a registered charity, has its central office at 2 Clockhouse Road, Farnborough, Hampshire GU14 7QY. It exists to campaign and raise awareness of anaphylaxis, to ensure that those affected have adequate guidance and treatments, and to promote research into severe allergies. Information is spread by means of leaflets, newsletters, videos, and a telephone helpline. The Campaign is run by an executive committee (largely comprising people directly affected by allergies) and is guided by some of the UK's leading medical experts in the allergy field.

9.2.4 The Food and Chemical Allergy Association

The Food and Chemical Allergy Association, based at 27 Ferringham Lane, Ferring, West Sussex BN12 5NB, came into being as a result of a letter sent to a daily newspaper in 1976 by its founder, Ellen Rothera. She had been ill for eight years and came to believe that food allergies due to a malfunctioning immune system were the root cause. She managed to stabilise her condition and make a recovery. Ellen's letter to the *Daily Express* was not only published, but given a leading position. As a result she was inundated with letters and telephone calls from people desperately seeking answers to their own medical conditions. A small group gathered for a meeting and formed an association, which set out to find doctors with knowledge of allergy, learn from them and continue in a self-help role. A committee was formed and a secretary appointed to answer all enquirers.

The FCAA continued in this manner for some years but eventually its role was changed to that of an advisory service. Today the association, still run by Ellen Rothera, seeks to provide information on all forms of allergy-induced illness and encourage people to look for the causes of their conditions instead of relying automatically on drugs to suppress the symptoms. The organisation provides general information on food allergy and chemical sensitivity and specific information on the more common forms of allergic reaction. It accepts that its approach to chronic illness, and the role that may sometimes be played by food, is not always accepted by the British Medical Association.

9.2.5 Action Against Allergy

Action Against Allergy is an independent charity founded in 1978 by Amelia Nathan Hill. She was chronically ill with migraine, stomach upsets, painful limbs and joints and other severe symptoms and her doctor, who could find no cause, thought she was being poisoned. After many years of searching, she undertook an elimination diet devised by British allergy pioneer Dr Richard Mackarness and found subsequently that her health improved dramatically.

Action Against Allergy, whose address is PO Box 278, Twickenham, Middlesex TW1 4QQ, adopts a wide definition of allergy, being convinced that its effects range from moderate symptoms to a severely debilitating chronic condition. AAA believes these can be triggered by a wide range of causes, including food, food additives, pollutants and chemicals.

AAA does not confine its help to those who become subscribers. The organisation offers information packs, advisory leaflets covering diet and allergy management, and quick reference sources of additive-free foods and non-allergenic products. Members, who pay £10 a year, receive an allergy newsletter and are able to use the AAA national talk-line. The charity also maintains a database of allergy clinics, specialists and dietitians.

9.2.6 Food Labelling Agenda

FLAG (Food Labelling Agenda) is a national consumer pressure organisation launched in June 1997 by a group of concerned food and health writers. The organisation campaigns for 'clear, comprehensive and meaningful labelling on all food and food products' and its first task in March 1998 was to deliver a petition calling for improved food labelling to Downing Street. It won support from a huge number of individuals and organisations, including those with interests in allergy, genetic engineering, infant feeding, heart disease, cancer, vegetarianism, eating disorders and green issues. The accurate labelling of potential allergens is one of FLAG's major concerns. The organisation is steered by Michelle Berriedale-Johnson and Sarah Stacey and their postal address is PO Box 25303, London NW5 1WN. A newsletter is produced for supporters.

9.3 Collaboration with governments

For many years, food allergy and intolerance had a poor public image. Despite the progress made in this field by a small number of dedicated scientists and physicians, allergy found itself on the fringes of medicine, dismissed altogether by some doctors, who regarded it as a convenient scapegoat for undiagnosed conditions that had other, unknown causes. People who claimed to suffer adverse reactions to food were accused of jumping on to an allergy bandwagon. Perhaps these accusations were justified in some cases, but many doctors threw out the baby with the bathwater. The cause was not helped by articles published by lightweight glossy magazines, which made extravagant claims about food allergy which plainly had little basis in science but were merely sensationalist.

However, all this began to change in late 1993 when five deaths triggered by allergic reactions to peanuts or nuts received widespread national publicity. Almost overnight, allergy became a serious issue. The death of my teenage daughter Sarah was reported by the *Mail on Sunday* in December 1993 – two days before an inquest was held – and subsequently by numerous other newspapers and television news programmes just before Christmas. Sarah's death caught the imagination of the media and the public generally. She was an attractive young hairdresser who had taken a rail trip to town one afternoon on her own and had treated herself to a slice of lemon meringue pie in her favourite restaurant. She had died a few hours later of anaphylactic shock, the most severe symptoms being severe bronchospasm and a dramatic fall in blood pressure. I am sure many people were deeply moved by the image of this innocent young girl being killed by something as innocuous as the peanut. Other deaths which happened within the same few months made it clear that Sarah's death was no freak incident. A 16-year-old Dorset schoolgirl, Michaela Mortimer, had died in early November after eating a biscuit containing peanuts at her school. Then there was Rachel O'Neill, a 26-year-old Suffolk woman who had died in equally shocking circumstances. Reporting all these tragedies, the *Mail on Sunday* article stated that Sue Nichol, the mother of a son with peanut allergy, was urgently trying to form a lobbying group of parents to get together to fight for improved food labelling.

Quite independently, other parents responded to the burst of media publicity by writing to the letters pages of national newspapers, calling for improved food labelling, and within a few weeks I had made contact with several of them, including Sue Nichol. About a dozen of us met in a flat off Baker Street, London, in January 1994 and this was the beginning of the Anaphylaxis Campaign. During those early days we had little knowledge of food labelling legislation and our approach was both naive and straightforward: we believed firmly that legislation was essential to ensure that peanuts or any other ingredients that could prove lethal for susceptible people must always be declared on the labels of prepacked food and on menus. Other people who were motivated by the publicity into joining the campaign clearly had similar objectives. Individually, these people began writing to Nicholas Soames, then Minister for Food at MAFF, and the weight of the correspondence that ended up on his desk began to take effect. It was clear to Soames and his officials that these people were not jumping on to the latest bandwagon, but were genuine people describing genuine symptoms.

Soames's reply to their letters made it clear that he took the issue seriously. He and his officials said they were impressed by the responsible tone of the letters: the correspondents were not fanatics but serious people who expressed themselves calmly and rationally.

Meanwhile the food industry was approaching Soames from another angle. Manufacturers, retailers and caterers could not escape the implications of the recent deaths. They were deeply alarmed by the fact that nuts could kill and were desperate for guidance. But at that stage, information was hard to come by. How much of a peanut could kill? Would a mere trace be dangerous? How many people were at risk?

By February 1994, the Anaphylaxis Campaign already had several hundred members or potential members, each one with a story to tell. Assuming the cases were all genuine – and most of them were – it was clear that the Anaphylaxis Campaign had a large storehouse of information. This file was given added credibility by Dr Rita Brown, consultant allergist at the Royal Berkshire Hospital, Reading, who provided the Anaphylaxis Campaign with evidence gleaned from her own work with nut allergy patients. A written report I compiled for MAFF reported her view that anaphylactic reactions are much more common than has been recorded and it is likely that some deaths from anaphylactic shock go undetected. She said that out of 663 consecutive new patients referred to her for allergy assessment, a total of 34 had suffered anaphylactic reactions to nuts, including peanuts.

During February, I heard on the grapevine that Soames was hoping to meet Anaphylaxis Campaign representatives so we could share our information with him, and this meeting, set up by my MP, Cranley Onslow, took place on 24 March in his office in Whitehall Place. We had been reliably informed that

Soames was an honourable man who would be unlikely to have a hidden agenda and indeed this proved to be the case. He said he was 'staggered' by the scale of the problem of potentially fatal food allergy and announced that he had already begun to launch an awareness campaign throughout the food industry. His officials at MAFF had met leading figures in the industry to discuss the dangers that certain foods posed for some people and information acquired was being passed down the food industry chain. But Soames said the labelling issue would have to be fought at European level and it would take time to change the EC labelling regime. And he made it clear that in his view, introducing tough labelling rules in catering establishments was not appropriate.

As a journalist, I have been taught to be suspicious of good intentions voiced by politicians, but Soames and his officials proved true to their word. Improvements in food labelling and in public awareness are there for all to see. That initial, fruitful meeting led to further talks with MAFF officials and industry representatives, a liaison that has continued through to the present day. More recently, support groups such as the Anaphylaxis Campaign and the Coeliac Society have provided high-quality input, through MAFF's consultative process, into the Codex Committee on Food Labelling, the international organisation that sets worldwide standards. During a long, laborious process, Codex agreed a list of foods known to cause hypersensitivity which should always be labelled, including peanuts, tree nuts, egg, milk, fish, shellfish, soya and gluten.

There is no doubt that the seriousness of nut allergy – and the high profile it now occupies – has been a major factor in leading the food industry, doctors and the public at large towards accepting the reality of allergies in general as a major cause of ill-health. The Anaphylaxis Campaign has played an important part in this process. But other support organisations, working tirelessly in the background, have continued to have an effect on government and public thinking.

In its very own sphere, far different from that of anaphylaxis, the Coeliac Society has been quietly active in keeping the needs of people with coeliac disease on the government's agenda. Apart from its work with MAFF on Codex food labelling matters, the society has been working to ensure that the presence or absence of gluten is declared on the labelling of foods marketed for babies aged 4–6 months. This is now law. In addition, the Food Labelling Regulations (1996) were amended in 1998 to require ingredients of foods identified as starch and modified starch to indicate their specific vegetable origin where they may contain gluten.

9.4 Collaboration with the food industry: retail and manufacturing

Soames's statement that any change in labelling legislation is a European matter is quite true. That prospect was an alarming one for campaigners who were concerned about food-induced anaphylaxis and the risks of allergy sufferers inadvertently coming into contact with lethal ingredients. Soames seemed to be implying that it would be several years before all prepacked food would be adequately labelled and therefore safe. A major point for discussion was the 25% rule governing compound ingredients. Under this European regulation, any ingredient which itself consists of more than one ingredient (e.g. the salami used on a pizza topping or a sponge finger on a trifle) does not have to have its component ingredients listed if it constitutes less than 25% of the finished food. Consequently, small amounts of an ingredient may be undeclared, quite legally.

But it soon became clear that major retailers and manufacturers were responding voluntarily to allergy issues, irrespective of the regulations. Weeks after the meeting with Soames, Britain's leading supermarket chains announced that their own-brand products would show nuts on the ingredient listing whenever they are present, regardless of the 25% exemption rule. Additional measures differed from company to company, but included the provision of nut-free lists for customers and a pledge that delicatessen products – and other items sold on in-store counters – would declare nuts on the counter tickets. Some companies have turned their attention to other allergenic ingredients, such as egg, milk, sesame, shellfish and soya, and better information is almost always available for people allergic to these products. Many manufacturers have adopted similar policies, particularly where peanuts and tree nuts are concerned.

The question of whether allergic customers should be given an extra warning – such as a coloured flash or symbol – has generated much debate. The views of individual customers differ on this, some people wanting a prominent warning about the presence of nuts, others preferring a statement guaranteeing that a product is nut-free. The official view of the Anaphylaxis Campaign – not necessarily shared by every member – is that the prime concern is to get all allergenic ingredients printed in the ingredient list. Although this is sometimes hard on those with poor eyesight, we feel that people should be able to rely on one simple, uniform system of getting information. Coloured flashes or symbols that differ from company to company may only serve to confuse, particularly when these are placed well away from the ingredient list. What may be helpful is an additional statement, CONTAINS NUTS, for example, placed immediately under the ingredient list.

The problems faced in supermarket in-store bakeries are sometimes viewed as insurmountable. Bakery staff make a large range of products including bread, cakes, shortbread, doughnuts, trifles and Danish pastries, and they use peanuts, nuts and seeds in a small but significant number of these products. Consequently most supermarkets display prominent signs discouraging customers with nut or seed allergies from buying *any* food from their in-store bakeries because of the possibility of cross-contamination. These signs cause intense anger among allergic customers. In an attempt to understand the problems, representatives of the Anaphylaxis Campaign spent an afternoon in a typical in-store bakery and discovered that the risks are real. Lying at the root of the problem are two main factors: limited space and the human element. We noted that staff operating within a small area work quickly under pressure to meet daily deadlines. Although staff employed by that particular company are trained in allergy issues, mistakes can happen. Seeds or tiny chunks of nut occasionally wander, carried, perhaps, on a baking tray or some other container. The chance of a nut or a seed ending up on a product bought by an allergic customer is very remote indeed – but the risk does exist.

In response to the exasperation expressed by members, the Anaphylaxis Campaign has raised the issue on many occasions during discussions with retail companies. We believe there may be some room for manoeuvre. Supermarkets are probably right to discourage people with nut allergies from eating cakes or pastries bought in their in-store bakeries, but they might reduce risk where bread-making is concerned. Managers might look at their operations and consider whether it is possible to dedicate their bread-making area as a nut-free zone. Instead, most of them effectively put in-store bakery products out of bounds for people with severe allergies. I will be returning later to the general problems of cross-contamination and disclaimer labelling.

Because public attention had been focused on peanuts and nuts – those foods most commonly implicated in severe reactions – there is a danger that food companies may overlook problems presented by other ingredients. Those occasionally implicated in the UK in serious incidents include milk, egg, sesame seeds and shellfish. The most frequent problems appear to be caused by minute amounts of milk products, quite legally undeclared under the 25% rule. A few examples are as follows. A 12-year-old boy with severe milk allergy suffered a moderately serious reaction when he ate an individual apple pie manufactured by a major UK company. Quite openly, the company said milk was present in a minute quantity – believed to be 0.006% of the finished product. The same boy suffered a reaction when he ate a cereal product made by a major manufacturer. Two other children reacted to small quantities of milk at around the same time – one to a sausage, the other to a crisp-type snack. In all these cases, the milk products were quite legally undeclared. In all cases, it was heartening that the companies concerned took the problems seriously.

Consumers with allergies should be encouraged to enquire about the free booklets produced by some retailers and manufacturers. These list products free from certain ingredients such as milk, egg, soya, gluten and shellfish, as well as nuts.

So much for what has been achieved – but what about the mistakes that occasionally occur? What happens when a nut chocolate bar ends up on the shelves bearing the wrong wrapper? Or a customer opens a box of chocolate raisins to find the peanut variety inside? A crisis management expert who addressed a food industry conference in 1996 stated categorically that in such cases, the best course open to food companies was to come clean. Sweeping such mistakes under the carpet, he warned, would only lead to disaster. Fortunately, this is the thinking adopted by most – if not all – food companies when something goes wrong. During a four-month period, these were just a few of the crises that occurred:

- A nougat bar withdrawn from sale because it contained egg undeclared in the ingredient list.
- Packs of toffee yoghurt found to contain hazelnut.
- Frozen chicken fillets found to contain a small amount of egg, quite legally undeclared on the labelling because of the minute quantity used.
- Some packs of a toffee ice cream dessert thought to contain hazelnut.
- Some packs of a well-known cereal thought to contain traces of peanut.
- Packs of chocolate raisins found to contain the peanut variety.

In all cases, the companies concerned liaised with the Anaphylaxis Campaign and took action. This action varied, but in the most serious cases involved a full product recall, extensive warnings placed in the national press and a mailshot to members of the Anaphylaxis Campaign. The Campaign's mailshot system is operated where there is a significant risk to people from a food product. The usual procedure is for the company concerned to use address labels supplied by us to send a letter to each member. The company makes a commitment not to use members' details for any other purpose. We find that our members are grateful to companies that make use of this service. The British Allergy Foundation has recently adopted its own early warning system.

There is one further, important development which has huge implications for the food industry and for allergy sufferers. In 1997, research into the allergenicity of peanut oil was completed in Southampton and published in the *British Medical Journal*. Under strict medical surveillance, 60 peanut-allergic adults were fed refined peanut oil and also unrefined peanut oil. As a result, six of them suffered allergic reactions to the crude oil, but these were only mild reactions. None reacted to the refined oil. The researchers conclude that refined peanut oil will not cause allergic reactions for the overwhelming majority of peanut allergic individuals. The research was funded by the London-based Seed Crushers and Oil Processors Association (SCOPA), which engaged in fruitful discussions with allergy groups. We believe the findings offer genuine reassurance to people who may have been anxious about the safety of refined peanut oil.

9.5 The use of disclaimers on food labels

In February 1994, one of the founder members of the Anaphylaxis Campaign gave a shrewd warning about the possible negative effects of any food labelling campaign: companies would begin to take the easy way out by printing disclaimer notices. A leading chocolate company had already begun to include a warning under the ingredient list of two of its brands stating 'May on rare occasions contain nut traces.' And dire warnings were given that this might conceivably catch on.

The prophecy has come true. Faced with the fact that tiny traces of certain allergenic ingredients may trigger a severe reaction in susceptible people, food companies have decided that in the case of some products it is impossible to guarantee safety. The warning 'may contain nut traces' is now a common feature of food packaging. The layman who is unfamiliar with the way food companies operate may wonder what it's all about – how do the nut traces get there and why bother to declare them? Most people who are allergic to nuts understand that the issue arises from what food technologists call 'cross-contamination' – where a product going down the production line inadvertently comes into contact with something that went down the same line previously. If food companies are willing to come clean about this, surely people with allergies should be grateful for any warning that protects them. The truth is that some are indeed grateful, but many have become exasperated by the increasing prevalence of these labels – which they suspect may be a cop-out.

A key question is: How likely is it that someone with nut allergy will come across a particle of nut on, say, a spoonful of breakfast cereal that is supposed to be nut-free? The answer is that it is probably very unlikely indeed, but it does occasionally happen. A young boy visiting a football match with his grandfather decided at half time to have his usual treat – a milk chocolate bar that he had eaten many times before. He suffered a mild allergic reaction. Looking at the label he saw the warning in small print: 'May on rare occasions contain nut traces.' This kind of incident may be rare, but the risks have rung alarm bells in the food industry.

To seek to understand the issue better, representatives of allergy groups, including the Anaphylaxis Campaign, have spent many hours visiting food companies. The average food production line is a complex affair: an arrangement of chutes, conveyor belts, vats, hoppers and tubes. At various points there are nooks and crannies. A thorough clean-up and other stringent measures will minimise risks, but – and this is where the ultra-cautious legal people usually have their say – no guarantees can be given that tiny nut traces will not remain.

Many allergic customers accept that 'may contain' labels are a necessary evil in some cases. They would rather see a warning than be put at risk. But the growing prevalence of disclaimers has led many people to believe that some companies are using them as a substitute for Good Manufacturing Practice because labels are cheaper and less bother than tightening up their act.

In 1997, the Labour government's Food Safety Minister, Jeff Rooker, challenged the use of what he called 'defensive labelling'. He wrote to manufacturers stating that this was acceptable only as a last resort after all methods of avoiding cross-contamination had been explored. And he told a food industry conference: 'I suspect that in many cases, these warnings are simply an alternative to adequate quality control.'

Most consumer groups probably agree with Dr Steve Taylor, an international expert in food science based in America, who told delegates attending a food conference in Leatherhead in 1996 that 'may contain' labels should be used 'judiciously' and only in situations where contamination is 'documented, uncontrollable, sporadic and potentially dangerous'. Many consumers go a stage further than this. They would expect that in the medium and long term, manufacturers will look closely at the option of opening segregated lines.

The British Retail Consortium, representing the supermarkets, and the Food and Drink Federation, for the manufacturers, have both issued guidance for their members on the identification and control of major food allergens, particularly nuts. The aims are to minimise risk, and consumers hope that if the result is indeed an improvement in manufacturing procedures, then there may be a reduced need for 'may contain' labels.

Significantly, both organisations say that 'may contain' labels should be a last resort of a series of assessments and should never be used as an alternative to Good Manufacturing Practices and relevant controls. And both advocate that dedicated lines should be regarded as an option.

Some may feel this is pie in the sky, but there are signs that a few companies are seriously considering the option of dedicated lines in order to remove 'may contain' labels. In the forefront is Kinnerton Confectionery, which began producing children's products on dedicated nut-free lines from spring 1999. The company took the revolutionary step of opening segregated lines and introduced far-reaching measures to avoid cross-contamination at its factory in Fakenham, Norfolk.

Kinnerton had decided that 'may contain' labels were an unsatisfactory solution to the problem of cross-contamination and that bolder measures were needed. As the factory wound down just before Easter 1999, the site was closed, production was stopped and operators took a three-week holiday. Then every production line except one was moved so that manufacture could be separated between lines containing nuts and those destined to be nut-free. Pipework was re-routed, a partition wall was erected down the length of the factory area and other stringent measures were taken. Pie in the sky? Consumers are bound to ask: If one company can do it, why can't others?

9.6 The catering industry

So far I have concentrated on food sold in supermarkets, but it has to be accepted that the greater risks for people with severe allergies lie in catering establishments, where the owners are under no legal obligation to label allergenic ingredients. Most of the known deaths from food-related allergies have occurred when the food was eaten away from home. Although information on near-fatalities is largely anecdotal, it is almost certain that most of these incidents happen when food is eaten in hotels or restaurants or bought from a takeaway.

Dr Richard Pumphrey, of the North West Regional Immunology Service, studied 52 cases of fatal food-induced anaphylaxis and found in 75% of cases that the food that triggered the fatal reaction was either eaten out, bought from a takeaway or eaten at a party.

No example is typical, but the following account of a near-fatal reaction has particular force because two allergy sufferers were affected. A family went into a restaurant and made their choice from the menu. Because the two daughters were allergic to nuts, the waiter was questioned directly about the ingredients of the food they had selected. The waiter looked busy, harassed and a little annoyed, but he disappeared into the kitchen briefly and returned with an assurance that the food was nut-free. But the information proved to be inaccurate and what had started as a pleasant family evening out ended up with both girls being treated at their local hospital's accident and emergency department.

A far more tragic case involved a 13-year-old Aylesbury girl who was allergic to peanuts. A family friend offered to go to the local Chinese and collect a takeaway and the youngster ordered a portion of chips with curry sauce. When the friend returned, the girl took one bite of the chips coated in sauce and decided she didn't like it. It transpired that the chef had used peanut butter to make the sauce. The girl died of anaphylactic shock. The restaurant staff had never heard of peanut allergy and did not state on the menu that sauces were made with peanut butter.

In a more bizarre case, thankfully not tragic, a 14-year-old boy from Kent was eating in a restaurant with his family. The waiter was questioned carefully and when it came to dessert the ice cream packaging was brought to the family and all seemed fine. The boy was asked if he would like nuts on the top and he said yes, he had no problem with nuts. Unfortunately the nuts were coated with albumen and sugar to make them separate and crunchy, and the boy suffered a severe reaction for which he required an adrenaline injection. The boy is allergic to egg and had reacted to the albumen.

Caterers writing to the trade press have occasionally appeared uncooperative in their approach to allergies, plainly motivated by the extreme difficulties involved in giving absolute guarantees that meals are free of nut traces. The view of some is that it is not their problem, it is the customer's. To some extent I would agree. People known to be at risk from severe reactions must take responsibility for their own safety. They must have their allergy assessed by an NHS specialist or a GP with an interest in allergy, and must always carry their prescribed medication so they can treat themselves in an emergency. They must learn to be direct with waiters, even if it means stating that a certain food could kill them, and learn to say no if they are not confident a particular dish is safe. It might also be argued that they avoid high-risk situations, such as eating out in Oriental restaurants where nut products are used liberally in the cooking. But they deserve assistance and understanding from catering staff. And there have been many examples where caterers have adopted sensible policies to minimise risks.

In 1997, the Ministry of Agriculture, Fisheries and Food produced detailed written guidance on allergies for caterers, but many restaurants had taken voluntary action long before then. It was in a Debenhams restaurant that my daughter ate her lethal dessert, and Debenhams reacted positively, immediately and, indeed, compassionately. The company introduced a system whereby customers with allergies can find out exactly what they are about to eat. Prominent signs invite customers to enquire about any dish on the menu, and once such an enquiry is made, staff produce a comprehensive product information book. This lists every meal, and every ingredient on the menu.

Chartwell, the school catering company, has devised a code of practice for its catering managers entitled 'A positive approach to managing food allergies in educational establishments'. This eight-page document explains in depth what anaphylaxis is and what causes it. More importantly, it outlines a series of steps that managers are expected to take, including collecting information from students or parents, recording this information and reviewing it regularly, communicating thoroughly with all staff, and identifying and training key members who will be responsible for having a full knowledge of recipes and ingredient information.

The MAFF initiative was launched in November 1997 by Jeff Rooker, who announced that an information pack would be sent to 175 000 catering establishments across the UK, including restaurants, cafés, hotels, pubs and takeaways. The pack contained a bright blue and yellow poster, information booklet, telephone sticker with emergency protocol, and 'allergy aware' sticker to display in the window. The pack gave clear messages that tiny traces of certain foods can kill, catering staff must give accurate information, and there must be positive measures taken to avoid cross-contamination. Caterers were urged: Think before using nut and seed oils, salad dressings and seafood sauces; don't let nuts, seeds and shellfish touch food that shouldn't have those ingredients; clean your hands, work surfaces and utensils after handling nuts, shellfish and seeds; think before cooking with oils that have been used to cook other foods.

Unfortunately, a follow-up by MAFF found that its project had been only moderately successful; some catering outlets actually displayed the material, others did not. The risks for allergic customers increase dramatically when they reach their teenage years and 'fly the coop'. New-found social freedom means they are having meals out with friends; they are often reluctant to follow advice; and their social environment (e.g. pubs, clubs and restaurants) may bring them into contact with peanuts and nuts more and more. The risks are manageable, but many people have a resistance to confronting the problem.

Returning to the cold reality, we have the case of a young woman with nut allergy who died after going out for a meal with colleagues in a top hotel in 1995. She was unaware that the butter contained nuts. We have the case of the 18-year-old economics student who died when she suffered an overwhelming allergic reaction to nuts in a dessert during her first night at Cambridge University in 1998. We have the 19-year-old man who died the same year after eating nuts in an Indian takeaway. And we have the well-publicised case of the athlete Ross Baillie, who died in June 1999 after taking a few bites from a coronation chicken sandwich containing nuts.

Such tragedies are rare and may become rarer if allergy sufferers are able to benefit from what amounts to a three-way partnership. First and foremost, it is vital that the sufferer seeks medical advice; second, the medical profession must ensure that such people are able to carry the appropriate medication so they can protect themselves; and third, sufferers have the right to expect the cooperation of caterers.

Rooker said in the House of Commons in June 1999 that he was now considering a Code of Practice for catering establishments which would seek to put allergy awareness into a tight framework. This is currently being discussed by the new Food Standards Agency.

9.7 Coeliac disease

Coeliac disease deserves a special mention. When discussing this lifelong condition, it is essential that it is not confused with classic food allergy. Allergic reactions usually occur within seconds or minutes of contact with the offending food, or occasionally within hours, whereas coeliac disease rarely causes such an acute, immediate reaction. This is an important distinction, because there is a danger that food companies may place gluten traces in the same danger category as nut traces, with the result that they fall to the temptation of adopting 'may contain' labels. There have certainly been cases where a trace of gluten makes a coeliac unwell, but such cases are not fatal.

The symptoms of coeliac disease are very specific. In babies they consist of pale, foul-smelling stools, wind, bloating and poor growth. These symptoms usually develop a few weeks after cereals are introduced into the diet. In those cases where coeliac disease begins as an adult, the symptoms are diarrhoea, pain, bloating, weight loss, malaise and weakness. In rare cases, constipation may be the main symptom. Nearly 90% of newly diagnosed coeliacs are over 16 and babies nowadays represent only 5% each year. Coeliac disease can severely reduce the quality of life for those affected, and it is vital that people affected avoid the specific food that causes it: wheat and related grains.

Coeliac disease used to be considered rare but more doctors are beginning to recognise the disorder in their patients. The Coeliac Society believes the average prevalence could be as high as one in 300 people in the UK and Europe. The only treatment is to return the intestine to normal by means of a strict gluten-free diet. Sometimes vitamin or mineral supplements are required to start with, but these should not be taken without medical supervision. To avoid serious complications of the condition, a strict gluten-free diet is necessary for life.

It is clear from the above that coeliac disease is an important issue and one that needs to remain on the agenda for the UK food industry.

9.8 Research into allergy and intolerance

Much progress has been made in recent years in understanding the mechanisms of allergy, but our knowledge is far from complete. Despite good work done in the UK and the United States and elsewhere, it is still uncertain how and why some people become allergic to certain foods and substances. As far as the allergy sufferer is concerned, all he or she can really do is try to avoid the offending food, scrupulously carry around prescribed medication, devise an action plan for when things go wrong – and wait and hope for better treatments to be developed. Further research is the customer's best hope.

The fact that our knowledge is incomplete poses obvious difficulties for support groups, which rely on information provided by the medical community. This information may have to be modified from time to time and it may even change altogether. Key messages conveyed by support groups one year may be overturned the next, so that advice offered in good faith may later prove to have been unwise or even dangerous. As a purely theoretical example, no one knows for sure what effect minuscule traces of peanut may have on the development of allergy. Traces may sensitise, or they may protect against sensitisation; occasional traces may sustain sensitivity in a person already sensitised, or they may lead to tolerance; occasional traces may protect a person from a severe reaction, or they may build up to a catastrophic reaction. How can we be sure that by advocating total avoidance of allergens we are putting forward the correct messages?

Other questions occupy the forefront of our thinking. Why is nut allergy particularly dangerous? Are there other foods that may be climbing up the allergy table? Today peanuts – what next? Might there be a way to identify 'high-risk' allergy patients early in life? If this were possible, it would release an intolerable burden from those who think they are at risk of a severe reaction, but may not be. Should everyone with peanut allergy be offered adrenaline? Or are those who advocate adrenaline for all nut allergy patients generating needless complications?

Perhaps what we are looking for is a clinical test to identify patients in the high-risk group. It would also be valuable to know what factors increase a patient's vulnerability at any particular time. We know that the patient's general state of health may be important, and we suspect alcohol consumption may play a part in some cases. But there may be other factors coming into play.

Allergy support groups strongly advocate more basic research which will seek to unravel these mysteries and make the risk of anaphylaxis more manageable.

9.9 Summary

Food scares come and go. Some issues, such as E.coli 0157, are genuinely serious; others, such as salmonella in eggs, may have been seriously overplayed in the media. Headlines such as 'Killer nuts' may suggest an over-reaction on a slow news day, but they should not deflect us from the probability that, unless science develops a wonder cure, allergies are here to stay. The food industry seems to have grasped that fact, and has shown no sign of relegating allergy to the status of 'two-day wonder'.

What focuses people's minds is the extraordinary fact that tiny traces of certain foods – nutritious for the majority – can be very serious indeed and even kill highly susceptible people. It might have been tempting during the early days for food companies to have ignored the minority and catered for the majority. But that minority is probably not as small as people had realised. Even life-threatening allergic reactions appear to be happening more and more frequently. Where a child is believed to be at risk, food companies must surely consider that child's wider 'support network' – parents, brothers, sisters, grandparents, uncles, aunts, friends, schools. The list of potential customers grows.

Fortunately most food companies – partly motivated by self-preservation but also out of genuine concern – have expended large amounts of energy and money into proving they are serious. Overall, customers must find it heartening that their concerns have been taken seriously and, one would hope, their lives have been made safer.

The major issue of safe eating out will certainly continue to occupy the attention of support group leaders. Nicholas Soames stated in 1994 that legislation was not appropriate to bring caterers into line, and he is probably right. Equally, I believe customers must think hard before rushing into litigation. Quite apart from the fact that it is costly, a get-tough attitude would no doubt lead to a situation where caterers would move into defensive mode. Customers would see a sign on every café door telling them: 'If you're allergic to nuts, don't eat here.' Nevertheless, it must be argued that five years is long enough for caterers to have understood the issues and made some progress, and anyone showing a despicable lack of care would no doubt deserve the consequences.

The equally important issues of cross-contamination and defensive labelling will also run and run. In 1997 the Anaphylaxis Campaign told its members that it was working to ensure that the food industry would adopt 'a sensible approach' to cross-contamination and defensive labelling. Many of our members believe we have failed in this objective. Letters expressing anger and dismay at the growing number of 'may contain' labels make up well over half of our mail.

Responsible food companies have said they take this issue seriously and will adopt defensive labelling only as a last resort. Certainly the good practice guidelines issued by the British Retail Consortium (BRC) and the FDF offer real cause for optimism. But good intentions will be of value only if they are translated into action. In addition, they are good short- or medium-term solutions, but the Anaphylaxis Campaign believes that in the long term, food manufacturers will inevitably be looking at the option of opening segregated lines.

The Anaphylaxis Campaign pledged in March 1999 that it would work to ensure that the issues of cross-contamination and defensive labelling remain high on the agenda and sought a commitment from the industry that the work within the FDF and BRC was not the end of the matter, but part of a continuing dialogue. The Campaign said it would like a pledge from all sections of the food industry that they are determined to remove the 'may contain' risk in the medium and long term.

But ultimately, we must face the fact that the real answers lie with medical science. We must admit with humility that our knowledge of allergy is in its infancy, and further research will be needed to help us solve the many mysteries surrounding allergy.

9.10 Sources of further information and advice

Anaphylaxis Campaign
PO Box 149, Fleet, Hampshire GU13 0FA
Tel. 01252 542029
www.anaphylaxis.org.uk

British Allergy Foundation
Deepdene House, 30 Bellegrove Road, Welling, Kent DA16 3PY
Allergy helpline: 020 8303 8583.

Coeliac Society
PO Box 220, High Wycombe, Bucks HP11 2HY
Tel. 01494 437278

Food and Chemical Allergy Association
27 Ferringham Lane, Ferring, West Sussex BN12 5NB

Action Against Allergy
PO Box 278, Twickenham, Middlesex TW1 4QQ
Tel. 020 8892 2711

Food Labelling Agenda
PO Box 25303, London NW5 1WN
Tel. 020 7837 1228.

Ministry of Agriculture, Fisheries and Food
Ergon House, London SW1P 3JR

Leatherhead Food Research Association
Randalls Road, Leatherhead, Surrey KT22 7RY

10

The epidemiology of adverse food reactions

A. Khakoo, G. Roberts, G. Lack, St Mary's Hospital, London

10.1 Introduction

Epidemiology is the study of the distribution of disease within populations. In collecting data about patterns of disease in a population, researchers are able not only to describe the scope of the problem in quantitative terms but also to further our understanding of the disease and its pathogenesis. Important clues may be uncovered about possible risk factors which may lead to the development of novel preventative or therapeutic strategies. An example of this in allergy is the high association between early exposure to bacterial infections and the absence of allergy in later childhood (Matricardi *et al.* 2000). This raises the question of whether we will be able to prevent allergies developing in children by alteration of the gastrointestinal flora and has stimulated research into DNA vaccines that use common bacterial DNA sequences to modulate the infant's immune system.

10.2 Methodological issues

10.2.1 Defining adverse food reactions

There are internationally agreed definitions for adverse food reactions, as have been discussed in Chapter 1. Unfortunately, terms such as 'food intolerance' are still used inconsistently. Thus the term 'intolerance' according to internationally agreed definitions is taken to mean physiological reactions to foods that do not have an immunological mechanism. However the term 'cows' milk protein intolerance' is often used to describe an immunological reaction to cows' milk that is non IgE-mediated (Host *et al.* 1997). It is not uncommon for authors to use several definitions of food allergy or food intolerance within a single publication. Any critical analysis of epidemiological studies must begin with a detailed understanding of the definitions used (for example, Table IV in Zeiger *et al.* 1999). This is of critical importance in comparative studies where like must be compared with like.

The international definitions of adverse food reactions exclude food aversion. Such aversions have a psychological origin and cannot be reproduced under objective conditions when the patient and observer are blinded to the identity of the food consumed. In population studies, up to a third of adults may report symptoms of adverse food reactions; however, double-blinded, placebo-controlled studies (Young *et al.* 1994) show that the majority of these are not reproducible. This discrepancy between real and perceived adverse food reactions is likely to be accounted for by the large number of subjects who have food aversion. The failure to differentiate between food aversion and reproducible adverse food reactions in studies relying on self-reporting (Bjornsson *et al.* 1996) or open challenges (Crespo *et al.* 1995) must be considered when prevalence data is assessed.

In this chapter we will focus on food allergies as this form of adverse food reaction is the best studied and associated with the highest morbidity. Additionally, we will describe adverse reactions to food additives. Although any food can cause allergic reactions, most reactions are caused by a limited range of foods. In infants and young children, the commonly implicated foods are cows' milk, egg, soy, wheat and peanuts. For older children, teenagers and adults, foods such as fish and shellfish are also a significant problem.

10.2.2 Diagnosing adverse food reactions

Different studies vary considerably in their working diagnostic criteria for food allergy. This has an important influence on the resultant measurement of prevalence and incidence in a population. In looking at IgE-mediated allergic problems, there are three levels of diagnostic criteria: (1) questionnaire-based histories, (2) specific IgE and/or skinprick testing and (3) food challenges (see Chapter 3). If, for example, we compare two population studies defining the prevalence of cows' milk allergy, one using skin testing and the other questionnaire-derived data, a higher prevalence will emerge in the latter study design. Double-blinded, placebo-controlled food challenges represent the gold standard but can not be practically used in large population-based studies where a combination of skinprick testing and questionnaire-based histories is more applicable.

10.2.3 Measuring the frequency of adverse food reactions and relating this to the natural history

There are a number of ways of measuring the degree to which a population is affected by a disease process such as food allergy. The best approach depends on the question being asked. Investigators usually measure either the incidence or the prevalence. The incidence is the number of new cases of adverse food reactions developing over a specified time. This is a useful measure when studying causality and possible preventative strategies but it gives little idea of the proportion of the population affected by the problem. The prevalence is the proportion of a specified population who suffer from adverse food reactions at a particular time. This figure is useful for gauging the health burden imposed by the problem in a population. Prevalence rates reflect two dynamic processes: the acquisition of new cases of allergy in a population (incidence) and the simultaneous loss of allergy in that population (either due to death or clinical remission). Therefore a static prevalence between two time points may fail to reveal high incidence and resolution rates that negate each other.

Many studies quote cumulative incidence which does not reflect the resolution of the adverse food reactions. While this indicates the proportion of the population who will have suffered allergy over a defined time period, it tends to overestimate the magnitude of the problem at a given time point which is better reflected by the point or period prevalence. The natural history of food allergy must be taken into account if we are sensibly to interpret incidence and prevalence rates in a population. For example, cows' milk allergy mainly presents in the first year of life and is almost entirely outgrown over the following four years (Figure 10.1). Therefore its incidence peaks by one year of age decreasing to almost zero at five years. The prevalence of cows' milk allergy will increase until a year of age after which it also starts to decrease as the remission rate exceeds the incidence rate. Cumulative incidence meanwhile will increase to a plateau at one to two years of age (Dean 1997).

If prevalence or incidence rates are being compared across two populations, it is important that the two populations have similar demographics. Thus the two populations may need to be standardised with respect to male–female ratio and the age structure of the populations, as both these variables significantly affect incidence and prevalence rates for food allergy. Care is also required when

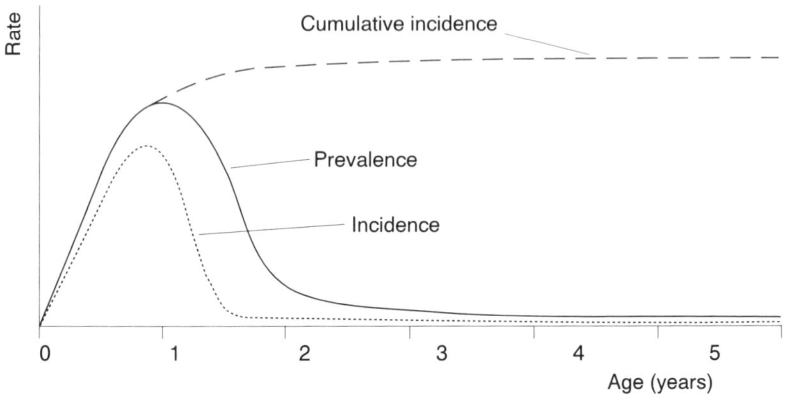

Figure 10.1 Natural history of cows' milk allergy.

prevalence rates are compared over time between different age groups in a population because of the cohort effect. If a higher rate of food allergy is found in infancy than adults, there are two potential explanations: firstly, this may reflect high remission rates of the problem or secondly it may be explained by a cohort effect in that the cohort of infants born in the 1990s may have a higher incidence of food allergy compared to the cohort of infants born several decades ago.

10.2.4 Implications of study design

In deciding on which study to use to estimate the prevalence of food allergy statistical, practical and financial constraints must be considered. The ideal sample would include all the individuals in the population but this is clearly impossible and our studies must be conducted on a subset of the total population. It is this down-sizing that leads to important methodological problems due to the selection procedures. The different types of study described below represent different selection procedures and give rise to different problems. It is impossible to obtain a subset that completely represents the entire population from which it is derived.

Case series

Many reports about food allergy have been based on personal series derived from general clinics or tertiary clinics. Such series are unable to provide any information about incidence and prevalence in a population as there is no known denominator associated with the data. Nevertheless, such series provide useful qualitative information about food allergies in different populations. Thus the fact that allergy to royal jelly is the most common cause of food allergy diagnosed in tertiary clinics in Hong Kong, but is never seen in European tertiary clinics is highly relevant (Leung *et al.* 1997). Furthermore, case series are useful in identifying novel problems. The fact that sesame seed allergy was rarely seen in European allergy clinics several decades ago but today represents an important component of the clinical case load suggests that this problem is increasing (Kanny *et al.* 1996).

Case series, however, are fraught with methodological problems, most notably bias. A bias is any error in the design or conduct of a study that results in a result other than the true one, due to systematic (though unintentional) skewing of the data. Bias may be introduced either in the selection of subjects or in the collection of information (Sackett 1979). A study looking at the association of soy allergy with cows' milk allergy (Zeiger *et al.* 1999) provides a good example of selection bias in a case series. The prevalence of soy allergy in one clinic was more than ten times greater than in the other three centres. This particular centre was a tertiary paediatric allergy clinic that saw a highly selected population, more likely to include children with multiple food allergies.

There is also a potential for information bias in case series because data are often collected retrospectively either directly from subjects or from their clinical notes. Patients often do not have a good recall of events, leading to a form of information bias called recall bias. A good example of recall bias is a birth cohort study in which mothers were asked about the duration of breast feeding at 11 and 47 months of age (Huttly *et al.* 1990). At 47 months there was only 70% agreement with data obtained from the same mothers at 11 months. Although case series do not provide robust epidemiological data, they provide a window through which current clinical experience may be viewed. They often form the initial basis of many hypotheses that can subsequently be tested in more definitive studies where cases and control subjects are compared.

Case-control studies

Case-control studies are a natural extension of case series having the added advantage that they provide control subjects with which the cases can be compared. Similar to case series, they provide no data on incidence and prevalence as here again the denominator remains unknown. They are, however, useful in the early testing of hypotheses that relate to associations and risk factors for food allergy. However, they make the assumption that all the differences between subjects and controls represent risk factors for the disease being investigated. In practice measured differences may be brought about by important biases in selection and information. Furthermore, confounding factors may occur where an apparent association between an exposure and an outcome is partially or entirely due to another associated exposure.

An example of a case-control study, is one looking at the aetiology of peanut allergy. It was concluded that children sensitised to peanut had a higher level of peanut exposure *in utero* due to higher maternal consumption (Frank *et al.* 1999). This result, which has not been confirmed in cohort studies, probably occurred because of recall bias as the mothers of infants with peanut allergy, are likely to have spent more time considering their consumption of peanuts during pregnancy prior to filling in the study questionnaire. Despite these potential problems, case-control studies represent a rapid way of providing important evidence about a hypothesis that can be later tested using a more definitive approach.

Cross-sectional studies

Cross-sectional studies are population-based studies within a defined geographical region. This approach considerably reduces the potential for selection bias. Furthermore they allow the point prevalence of the condition studied to be estimated. However, such studies by their nature afford a single glimpse of the population at one specific time point. Therefore no data can be derived about changes in incidence and natural history of the condition over time. Such studies allow us to identify risk factors for food allergies as the population contains both cases and control subjects. Cross-sectional studies involve large numbers of subjects and require considerable resources. They may also be affected by bias and confounding factors.

Mailed questionnaires are often used in cross-sectional studies but response rates can be very low, even after reminders are sent. In the High Wycombe population study of food intolerance (Young *et al.* 1994), replies were received from only 52.7% of subjects. It can be argued that responders are likely to differ substantially from the non-responders, introducing an important selection bias. Such problems may be reduced with door-to-door interviews but other problems emerge, for example, subjects out at work may escape interview. Telephone-based interviews are becoming increasingly common (Munoz-Furlong *et al.* 1989) but these exclude subjects without a telephone and those who choose to be ex-directory, which will bias the sample. Evidence from cross-sectional studies also provides useful data on the prevalence of disease in a population and highlights potential causal factors. Evidence collected using this approach must eventually be substantiated by the results of cohort studies to decide if it has been affected by selection or information bias.

Cohort studies

Cohort studies are less affected by the problems inherent in other approaches for the single reason that subjects are included and exposures recorded before the outcome has occurred. This eliminates a major source of bias. Cohort studies, unlike cross-sectional studies, are not subject to the cohort effect as all the participants are born over a specified narrow time period. Furthermore, one is able to estimate the incidence and remission rates as well as prevalence and thus obtain a more complete picture of the natural history of a disease. Such studies provide the best quantitative and qualitative description of food allergy within a population but make the highest demands on time and resources. Nevertheless, cohort studies are not completely immune to methodological problems. Selection bias may operate slowly over a longer period of time. At the start of the study, there is likely to be a loss of participants due to failure to enrol while others may become lost to follow-up during the study. The loss of potential subjects at enrolment and during follow-up is likely to introduce important selection bias.

Cohort studies are important in identifying risk factors for food allergy. This risk is usually quantified using odds ratios or relative risks. Confounding can still occur where a third factor may account for a perceived association between a particular exposure and an allergic outcome. Where such confounding variables are suspected and identified, their effects can be eliminated by the application of statistical methods such as logistic regression analysis. An example is the association seen between prolonged breast feeding and food allergy. This is not a real association as it is confounded by eczema; infants with eczema are deliberately breast fed for longer periods and eczema is a known risk factor for food allergy.

Although cohort studies have their limitations, they generally provide the best form of evidence concerning the prevalence and natural history of a disease within a population. They are well suited to the study of the natural history of food allergy. They also provide pointers to potential causal factors which can be subsequently tested within the context of a randomised interventional study where allocation of the exposure is random and not subject to known or unknown confounders.

Interventional studies

In many ways interventional studies are very similar to cohort studies except that the investigator is able to allocate the exposure artificially, preferably at random. The randomised, double-blinded, placebo-controlled study (RDBPC study), where exposure allocation is random and known to neither subject nor investigator, is the gold standard for generating evidence. Unfortunately, although the data generated by a good study are invaluable, these studies are difficult to set up for financial, practical and ethical reasons. It is only ethical to randomise subjects between two or more interventions if there is no evidence to suggest that one is more beneficial than the other. It may be difficult to recruit a sufficiently large and representative study population to have sufficient power to be able to arrive at definitive conclusions. Even once a study population has been recruited problems may occur with loss to follow-up because of a perceived failure of the active or control intervention.

Most interventional studies in food allergy have focused on maternal dietary exclusion during pregnancy and/or lactation as well as modification of the infant diet. In general they have been unsuccessful. A problem that pervades all such studies is that elimination of a food from the diet may not be achieved to a sufficient degree or at an early enough time point to ensure a successful intervention. These 'no difference found' studies become very difficult to interpret, leading to 'newer and better' interventional studies.

10.3 Commonly reported food allergies

10.3.1 Cows' milk

Introduction

Cows' milk is an important weaning food in many countries. In recent years it has become practically ubiquitous, being found in an increasing range of commercially produced foods (Sampson 1998). There is extensive cross-reactivity between milks of different species (Businco *et al.* 1995, Carroccio *et al.* 1999). Cows' milk is one of the first foods to enter an infant's diet and therefore is often the first to cause problems. Adverse reactions to cows' milk can be divided into two main groups, immunological (IgE or non-IgE mediated) or non-immunological (Host *et al.* 1997, Host and Halken 1998). This latter group is mainly due to lactase deficiency and may be difficult to differentiate clinically from non-IgE mediated cows' milk allergy (Host *et al.* 1997, Bruinjzeel-Koomen *et al.* 1995). Cows' milk allergy gives rise to a spectrum of disease from immediate symptoms ranging from urticaria to anaphylaxis (Goldman *et al.* 1963, Sampson *et al.* 1992) and late symptoms which may not develop for 24 to 48 hours. Most early reactors have specific IgE to cows' milk (Hill *et al.* 1988, Host and Halken 1990).

Prevalence and natural history

Adverse reactions to milk presents early, median age 4–8 weeks with a range of 1–52 weeks (Jakobson and Lindberg 1979, Host 1990, Schrander *et al.* 1993). In general, infants do not develop symptoms on their first exposure to cows' milk (Host 1990). Half react within a week of first exposure and three-quarters within four weeks (Host 1990, Jakobson and Lindberg 1979). Of note, a quarter of the children have their first symptoms while being exclusively breast fed (Host 1990).

Symptoms suggestive of cows' milk allergy or intolerance are relatively common and are seen in 2–15% of infants (Table 10.1). This variability is probably due to different diagnostic criteria, study design, geographical differences and different ages. Where subjects have been prospectively recruited and diagnosis is based on food challenge, the cumulative incidence has been found to be remarkably similar at 1.9–2.3% in the first three years of life (Table 10.1). Unfortunately few of these studies differentiate between the different types of adverse reactions or whether specific IgE is present. Where they do, 25–50% of infants develop symptoms within 1–3 hours (Schrander *et al.* 1993, Jakobson and Lindberg 1979, Host 1990).

The peak prevalence of cows' milk allergy is seen at 1–2 years of age. After this age, children start to lose their reactivity (Table 10.1). The remission rate in one study was 56% at 1 year, 77% at 2 years, 87% at 3 years and 92% at 5 and 10 years (Host 1990, Host 1997). The majority of children with persistent cows' milk allergy beyond 5 years of age have IgE to cows' milk (Host 1990, Host *et al.* 1997).

Within the adult population, adverse reactions to cows' milk are reported by 0.7% of the population but only 10% of these have specific IgE to cows' milk. The prevalence of IgE-mediated cows' milk allergy in adults is therefore extremely low at 0.07% and is probably due largely to persistent allergy from childhood (Bjornsson *et al.* 1996, Niestijl *et al.* 1994) (Table 10.1). Lactase deficiency is probably responsible for most adverse reactions to cows' milk in adults (Bruinjzeel-Koomen *et al.* 1995).

10.3.2 Soy

Introduction

Soy is a fairly recent addition to the Western diet although soybeans have been eaten for centuries in the Far East. The most commonly reported adverse reactions to soy are gastrointestinal symptoms, often as an enterocolitis syndrome or colitis (Powell 1978). Specific IgE to soy is not thought to be involved (Zeiger *et al.* 1999). Skin lesions are occasionally seen but IgE-mediated symptoms or anaphylaxis are extremely rare (Cantani and Lucenti 1997).

It had been thought that most cows' milk allergic children also reacted to soy (Cantani and Lucenti 1997). However, the early studies relied on the history, RAST tests or skinprick tests to make the diagnosis. When open or blinded challenges are used to diagnose both the cows' milk and soy allergy, 11–32% of children are found to react to both (Table 10.2). This variation in results between studies probably stems from the fact that they each enrolled different populations with different proportions of children with IgE and non-IgE mediated cows' milk allergy.

Prevalence and natural history

In general, adverse reactions to soy are first seen in the later part of infancy although when infants are exposed to soy at a young age, reactions can occur (Bruno *et al.* 1997). The prevalence of self-reporting of soy allergy is very low unless a selected, atopic population is studied (Table 10.3). In atopic children presenting to allergy clinics the prevalence is 1.2–3% when diagnosed by open challenge. In unselected childhood population, the prevalence is only 0–0.3%. Most of these children have positive specific IgE to soy and react within four hours. Where soy allergic children have been followed up, all have been found to outgrow their allergy within seven years (Bock 1982). This correlates with adult studies where the prevalence of soy allergy has been estimated as zero (Table 10.3). Soy may behave as an aeroallergen in both adults and children; in Barcelona, aerosolised soy from the port has been linked to epidemics of asthma (Navarro *et al.* 1993).

10.3.3 Peanuts and tree nuts

Introduction

Over the last few decades, peanuts have become a ubiquitous part of the Western diet as they are a versatile form of easily digested protein (Lucas 1979). In a study looking at the use of dietary manipulation to prevent the development of food allergy, all infants in the control group were exposed to whole peanuts by their second birthday (Zeiger *et al.* 1989); occult exposure probably occurs even earlier. Adverse reactions to peanuts and tree nuts are generally IgE mediated, occurring rapidly with subjects presenting with dermatological, respiratory and gastrointestinal manifestations (Hourihane *et al.* 1997). Peanuts and tree nuts are responsible for a third of all admissions with anaphylaxis (Bock 1992). Peanuts are part of the legume family, they are more closely related to peas, beans, soy and lentils than the tree nuts. It has been suggested that there is extensive cross-reactivity between peanut and tree nuts in terms of sensitisation but not clinical reactivity (Sampson and McCaskill 1985, Bernhisel-Broadbent and Sampson 1989). However a recent British survey suggested that 50% of people with peanut allergy have symptoms with other nuts (Loza and Brostoff 1995). For the legume family, most subjects show sensitisation to at least two members of the family but very few subjects are clinically allergic to more than one (Bernhisel-Broadbent and Sampson 1989).

Prevalence and natural history

Peanut and tree nut allergy generally presents in childhood (Sampson 1990, Kivity *et al.* 1994). The majority of children react to peanut on their first

Table 10.1 Epidemiology of adverse food reactions to cows' milk – key studies

Author, date	Type of study	Country	No. of subjects	Point prevalence	Cumulative incidence	Definition
Halpern *et al.* 1973	Prospective series	USA	1084	1.8%		History
Dean 1997	Population based, birth cohort	UK	1218	4.4% at 1y; 1.9% at 2y; 0.4% at 4y	5.1% to 4y	History
Arshad *et al.* 1993	Population based, birth cohort	UK	1174	1.8% at 2y		History
Burr and Merrett 1983	Population based, cross-sectional	UK	475	1% in adults		History
Young *et al.* 1994	Population based, cross-sectional	UK	18 880	2.7%, all age groups		History
Niestijl Jansen *et al.* 1994	Population based, cross-sectional	Holland	1483	0.7%, 18–70y		History
Bjornsson *et al.* 1996	Population based, cross-sectional	Sweden	1812	1% (sp IgE) 0.07% (sp IgE and symptoms) aged 20–44y		Sp IgE ± history
Kajosaari 1982	Population based, cross-sectional	Finland	802	2% at 1y 5% at 2y 2% at 3y 0% at 6y		Open challenge at home

Kivity *et al.* 1994	Retrospective series	Israel, recent onset symptoms to food	112	0%, 10–48y		Open or double-blinded challenges
Gerrard *et al.* 1973	Prospective series	Canada	787		7.5%	Open challenge at home
Schrander *et al.* 1993	Population based, birth cohort	Netherlands	1158		2.3% to 1y	Open challenge
Jakobson and Lindberg 1979	Population based, birth cohort	Sweden	1079		1.9% to 1y	Open challenge
Bock 1987	Population based, birth cohort	USA	480		15% to 3y (history) 2.2% to 3y (challenge)	History ± open/ double-blinded challenge
Host and Halken 1990	Population based, birth cohort	Denmark	1749		2.2% to 3y	Open or double-blinded challenges

Table 10.2 Combined cows' milk and soy allergy – key studies

Author, date	Country	Number of subjects	Basis of diagnosis of cows' milk allergy	Basis of diagnosis of soy allergy	Prevalence of soy allergy in children with proven cows' milk allergy	Comments
Kuitunen *et al.* 1975	UK	35	Open challenge	Open challenge	11%	Non-IgE, mean age 5m
Perkkio *et al.* 1981	Italy	103	Open challenge	Open challenge	11%	Non-IgE
Bardare *et al.* 1988	France	29	Open challenge	Open challenge	17%	Mixed
Paganus *et al.* 1992	Finland	19	Open challenge	Open challenge	32%	Mixed, mean age 11m
Zeiger *et al.* 1999	USA	93	Open or blinded food challenge	Open or blinded challenge	14%	Mainly IgE, mean age 19m

Table 10.3 Epidemiology of adverse food reactions to soy – key studies

Author, date	Type of study	Country	No. of subjects	Point prevalence	Cumulative incidence	Definition
Dean 1997	Population based, birth cohort	UK, unselected	1218	0% at 1y; 0% at 2y; 0% at 4y	0% to 4y	History
Young *et al*. 1994	Population based, cross-sectional	UK	18 880	0.3%, all age groups		History
Niestijl Jansen *et al*. 1994	Population based, cross-sectional	Holland	1483	0%, 18–70y		History
Bjornsson *et al*. 1996	Population based, cross-sectional	Sweden	1812	2% (sp IgE) 0% (sp IgE and symptoms) aged 20–44y		Sp IgE ± history
Giampietro *et al*. 1992	Prospective series	Italy, atopic children	317	3%, 1m to 10y		Open challenge
Kivity *et al*. 1994	Retrospective series	Israel, recent onset symptoms to foods	112	0%, 10–48y		Open or double-blinded challenges
Magnolfi *et al*. 1996	Prospective series	Italy, atopic children	704	21% by SPT, 1.1% by DBPCFC 1m to 18y		SPT± Double-blinded challenge
Sampson 1988	Prospective series of children with eczema	USA	204	5%		Double-blinded challenges
Bruno *et al*. 1997	Prospective series, multi-centre	Italy – infants with history suggestive of food allergy	505	1.2% at 6m to 14y		Double-blinded challenge
Bock 1987	Population based, birth cohort	USA (middle-class community)	480		2.2% to 3y (history) 0.4% to 3y (challenge)	History ± open/double-blinded challenge

Table 10.4 Epidemiology of adverse food reactions to peanut and tree nuts – key studies

Author, date	Type of study	Country	No of subjects	Point prevalence	Cumulative incidence	Definition
Burr and Merrett 1983	Population based, cross-sectional	UK	475	0% adults		History
Foucard 1991	Cross-sectional medical students	Sweden	1050	9%		History
Young *et al*. 1994	Population based, cross-sectional	UK	18 880	1.7% (all nuts) in all ages		History
Niestijl Jansen *et al*. 1994	Population based, cross-sectional	Holland	1483	0% adults		History
Emmett *et al*. 1999	Population based, cross-sectional	UK	46 252	0.61% (0–4y) 0.53% (15–44y) 0.30% (>44y)		History
Tariq *et al*. 1996	Population based, birth cohort	UK	1456	1.3% (SPT) at 4y 1.1% (history and SPT) at 4y		See left
Kajosaari 1982	Population based, cross-sectional	Finland	802	2% at 1y 1% at 2y 2% at 3y 0% at 6y		Open challenge at home

Munoz-Furlong *et al*. 1989	Population based, cross-sectional	USA	12 032	0.4% (0–17y) 0.7% (>17y)		History via telephone interview
Bjornsson *et al*. 1996	Population based, cross-sectional	Sweden	1812	3% (sp IgE) 0% (sp IgE and symptoms) aged 20–44y		Sp IgE ± history
Kivity *et al*. 1994	Retrospective series	Israel, recent onset symptoms with food	112	20%, 10–48y		Open or double-blinded challenges
Golding *et al*. 1998	Population based, birth cohort	UK	14 000		0.21% to 2y 0.31% to 4y 0.67% to 5y	Double-blinded food challenge
Bock 1987	Population based, birth cohort	USA	480		1.3% to 3y (history) 0.6% to 3y (challenge)	History ± open/double-blinded challenge

known exposure. Sensitisation must therefore be due to occult exposure (Hourihane *et al.* 1997). Reactions to peanuts and tree nuts are relatively common. Up to 2% of infants have histories of adverse reactions to peanuts with the highest prevalence figures being seen around four years of age (Table 10.4). Rates in adults appear to be lower unless a highly selected atypical population is studied (Foucard 1991). Where challenges are used to confirm the diagnosis, the prevalence figures drop to under 0.7% (Table 10.4). Historically, peanut and tree nut allergies have been considered to be lifelong problems (Sampson and Scanlon 1989, Bock and Atkin 1989). However, recently, evidence has been presented to suggest that at least some children outgrow their allergy by five years of age (Golding *et al.* 1998, Hourihane *et al.* 1998). This is supported by the evidence that the prevalence of peanut allergy is lower in adults (Table 10.4).

10.3.4 Fish and shellfish

Introduction

Fish is usually introduced relatively late into the infant diet and is therefore one of the less common infant food allergies. Shellfish usual enter the diet even later and adverse reactions to these are usually not seen until the teenage years or adulthood. Both fish and shell allergy are generally IgE mediated with a rapid onset of symptoms. Both have been implicated in anaphylaxis (Kemp *et al.* 1995, Yunginger *et al.* 1988, Bock 1992).

There is cross-reactivity between different species of fish – more at the immunological level than the clinical level. Sera from subjects with codfish allergy cross-reacts with proteins from other species but not with shrimp or milk (Hansen *et al.* 1997, de Martino *et al.* 1990, Helbling *et al.* 1999). Cross-reactivity at the immunological and clinical levels is also seen with shellfish (Castillo *et al.* 1994). It is unclear, though, to what extent there is clinical cross-reactivity between fish and shellfish.

Prevalence and natural history

Adverse reactions to fish are reported in less than 0.5% of young children (Table 10.5). An exception is one Finnish birth cohort study where 5% of infants reacted to fish on an open challenge at home (Kajosaari 1982). Fish allergy seems to increase with increasing age with up to 1.5% of adults having adverse reactions to fish by history (Table 10.5). However, in one study only 0.1% adults had both symptoms and specific IgE to fish (Bjornsson *et al.* 1996). Up to 2.1% of the adult population report adverse reactions to shellfish; no paediatric studies report significant shellfish allergy in children (Table 10.5). The natural history of fish allergy is unclear but there is one study that suggests that most children do not outgrow this problem (Bock 1982).

Table 10.5 Epidemiology of adverse reactions to fish and shellfish – key studies

Author, date	Type of study	Country	No. of subjects	Point prevalence	Cumulative incidence	Definition
Burr and Merrett 1983	Population based, cross-sectional	UK	475	1.5% (fish) 2.1% (shellfish) In adults		History
Bock 1987	Population based, birth cohort	USA	480		None to 3y	History
Young *et al*. 1994	Population based, cross-sectional	UK	18 880	2.9% (fish and shellfish) all ages		History
Niestijl Jansen *et al*. 1994	Population based, cross-sectional	Holland	1483	0% in adults		History
Dean 1997	Population based, birth cohort	UK	1218	0.2% at 2y 0.1% at 4y	0.2% to 4y	History
Bjornsson *et al*. 1996	Population based, cross-sectional	Sweden	1812	0.3% (sp IgE) 0.1% (sp IgE and symptoms) aged 20–44y		Sp IgE ± history
Rance *et al*. 1999	Prospective series, children with food allergy	France	703	0% <1y 3.3% 1–3y 8.5% 3–6y 17.5% 6–15y		Open or blinded food challenge
Kajosaari 1982	Population based, cross-sectional	Finland	802	5% at 1y 2% at 2y 3% at 3y 1% at 6y		Open challenge at home
Kivity *et al*. 1994	Retrospective series	Israel, subjects with recent onset allergic symptoms to food	112	None, 10–48y		Open or double-blinded challenges

10.3.5 Egg

Introduction

Egg is an early weaning food in diets worldwide. Although chicken egg is the predominant form in the developed world, in other countries eggs from other animals are commonly eaten, but epidemiology for these is lacking. Adverse food reactions to egg are acute and IgE mediated, implicated in cardiorespiratory, gastro-intestinal and skin (including exacerbation of eczema) reactions (Burks *et al.* 1998). Egg proteins may cause IgE-mediated occupational asthma in egg-processing workers (Bernstein *et al.* 1987). The most allergenic part is the egg white, but allergens also reside in the yolk. The eggs from other birds such as turkey, duck and goose contain the same major allergens as are found in chicken egg (Langeland 1983). Egg allergy is a marker for peanut allergy (Dean 1997), and appears to be the food most associated with later onset asthma/allergic rhinitis with up to one in three children developing skin test reactivity to aeroallergens by the age of four years (Dean 1997, Nickel *et al.* 1997). Key epidemiological studies are shown in Table 10.6.

Prevalence and natural history

There is no evidence of sensitisation to egg at birth (Kulig *et al.* 1999), and the majority (85–90%) of clinical reactions occur in the first 2–3 years of life while 55–88% of reactions occur on the first known exposure to egg (Ford and Taylor 1982, Langeland 1983). The incidence of clinical egg allergy varies with the definition used. It ranges from 2.4% in children up to four years old if the definition is based on history of a clinical reaction (Dean 1997) to 0.6% in children up to three years old if the more rigorous definition based on DBPCFC is used (Bock 1987). However, a study using open egg challenge at home as an endpoint suggests a peak prevalence of egg allergy of 9% at three years of age (Kajosaari 1982). No new case of egg allergy after the age of ten years was found in a study of 112 patients with food allergy presenting to an allergy clinic, supporting the notion that egg causes allergy only in the early years (Kivity *et al.* 1994). The rate of positive skinprick or specific IgE to egg is 1.5–4 times greater than clinical reactivity (Kjellmann *et al.* 1988, Dean 1997). In the atopic population the risk of egg allergy is increased by up to 3–5 times (Ratner and Uuntract 1952, Hill *et al.* 1997).

Up to 75% of children with egg allergy can tolerate egg by seven years of age (Kjellmann *et al.* 1988) and in one birth cohort study the mean age of developing tolerance was 19 months (range 12–42 months) (Kjellmann *et al.* 1988). Resolution is less likely to occur in those children with cardiorespiratory reactions to egg (Ford and Taylor 1982).

10.3.6 Wheat

Introduction

Wheat is another food consumed worldwide. However, there are geographical variations in consumption patterns and age of introduction into the diet; for

Table 10.6 Epidemiology of adverse food reactions to eggs – key studies

Author, date	Type of study	Country	No. of subjects	Point prevalence	Cumulative incidence	Definition
Young *et al*. 1994	Population based, cross-sectional	UK	18 880	2.3% at > 6 months old		History
Dean 1997	Population based, birth cohort	UK	1218	1.6% at 1y 1.4% at 2 y 0.6% at 4y	2.4% to 4y	History
Varjonen *et al*. 1992	Atopic population, cross-sectional	Finland	434	0.5% at 15–16y		History and SPT
Dean 1997	Population based, birth cohort	UK	981	0.8% at 4y		SPT
Hill *et al*. 1997	Population based, birth cohort	Australia	332		3.2% to 2y	SPT
Hill *et al*. 1997	Atopic population, birth cohort	Australia	620		16.4% to 2y	SPT
Bjornsson *et al*. 1996	Population-based, cross-sectional	Sweden	1397	0.8% (sp IgE); 0.2% (sp IgE and history) at 20–44y		History ± sp IgE
Kulig *et al*. 1999	Population based, birth cohort	Germany	4082	0% at birth 6.3% at 1y 4.5% at 3y 4.5% at 6y	13.4% to 6y	Sp IgE
Kajosaari 1982	Population based, cross-sectional	Finland	802	6% at 1y 7% at 2y 9% at 3y 1% at 6y		Open challenge at home
Bock 1987	Population based, birth cohort	USA	480		2.3% to 3y (history);0.6% to 3y (challenge)	Open history ± double-blinded food challenges

example in some Asian countries such as Japan it is consumed earlier and in larger amounts (Hill *et al.* 1997). Wheat causes a range of allergies, either IgE-mediated causing acute hypersensitivity and delayed eczema reactions seen mainly in infants, or a cell-mediated immunological inflammation of the small bowel, coeliac disease, seen in children and adults. Key epidemiological studies are shown in Table 10.7.

Prevalence and natural history

IgE-mediated wheat allergy is uncommon with a cumulative incidence of 1–2% in the first six years of life. IgE-mediated wheat allergy may occur less commonly in later childhood or adulthood (Kivity *et al.* 1994), and in the latter is also responsible for the occupational disease, baker's asthma. In young children, IgE-mediated wheat allergy causes predominantly mild reactions, but in adults wheat may be responsible for up to 63% of cases of exercise-induced food-related anaphylaxis (Guinnepain *et al.* 1996). Specific IgE data probably overestimate clinical reactivity by 2–3 times (Kulig *et al.* 1999).

The true prevalence of coeliac disease is unknown. The conventional view has been that prevalence in the UK is around 1:1500, but more recent data from the UK (Hins *et al.* 1999) and Italy (Catassi *et al.* 1994) suggests that 1:300 of the population in Europe may have some form of gluten sensitivity. In older studies coeliac disease was more common in the first two years of life, but the average age of diagnosis in childhood is now increasing, being five years of age in Finland (Ferguson 1999). This may be due to the later introduction of gluten into infant diet which delays the onset of clinical disease (Anderson 1992), and presentation in later childhood and adults who have previously clinically tolerated wheat is becoming more common.

Patients with coeliac disease need to pursue a lifelong gluten-free diet, and they also need to avoid rye, barley and possibly oats. There is not much good epidemiological data on the natural history of IgE-mediated wheat allergy, although in a select population of subjects aged 3–18 years with eczema and wheat allergy, 33% became tolerant within 1–2 years (Sampson and Scanlon 1989).

10.3.7 Fruits and vegetables

Introduction

Vegetables and fruits are staple foods in diets worldwide although the types of vegetables and fruits consumed vary widely. It is therefore not surprising that considerable geographical variations exist in respect of adverse reactions to specific fruits and vegetables. Vegetables, and more particularly fruit, may cause adverse reactions that are either IgE-mediated which most often have their onset after the first few years of life (in contrast to many other foods), or occur via other mechanisms, typically in early childhood. Key epidemiological studies are shown in Table 10.8.

Table 10.7 Epidemiology of adverse food reactions to wheat – key studies

Author, date	Type of study	Country	No. of subjects	Point prevalence	Cumulative incidence	Definition
Young *et al*. 1994	Population based, cross-sectional	UK	18 880	0.9% at > 6 months old		History
Dean 1997	Population based, birth cohort	UK	1218	0.5% at 1y 0.5% at 2y 0.2% at 4y	0.9% to 4y	History
Bjornsson *et al*. 1996	Population-based, cross-sectional	Sweden	1397	3% (sp IgE); 0% (sp IgE and history) at 20–44y		Sp IgE ± history
Kulig *et al*. 1999	Population based, birth cohort	Germany	4082	0% at birth 0.8% at 1y 2% at 3y 4% at 6y	5% to 6y	Sp IgE
Kajosaari 1982	Population based, cross-sectional	Finland	802	1% at 1y 1% at 2y 0% at 3y 0% at 6y		Open challenge at home
Bock 1987	Population based, birth cohort	USA	480		0.8% to 3y (history); 0.2% to 3y (challenge)	Open history ± double-blinded food challenges

Table 10.8 Epidemiology of adverse food reactions to fruit and vegetables – key studies

Author, date	Type of study	Country	No. of subjects	Point prevalence	Cumulative incidence	Comments
Foucard 1991	Cross-sectional, medical students	Sweden	1050	7% apple/related fruits		History
Young *et al*. 1994	Population based, cross-sectional	UK	18 880	3.5% citrus 1.2–1.3% tomatoes 1.0% non-citrus 0.5% vegetables at > 6 months old		History
Niestijl Jansen *et al*. 1994	Population based, cross-sectional	Holland	1483	2.2% vegetables 1.6% fruit at 18–70y		History
Saarinen and Kajosaari 1980	Population based, birth cohort	Finland	145		Citrus fruits 13.1% to 3y	History
Varjonen *et al*. 1992	Atopic population, cross-sectional	Finland	416	6.9% apple 3.8% carrot 1.4% celery 1.2% paprika 0.2% orange at 15–16y		History and SPT
Saarinen and Kajosaari 1980	Population based, birth cohort	Finland	145		Citrus fruits 3.4% to 3y	Open challenge at home
Kajosaari 1982	Population based, cross-sectional	Finland	802	Citrus fruits 5% at 1y 2% at 6y		Open challenge at home
Bock 1987	Population based, birth cohort	USA	480		Fruits and fruit juices 12% to 3y	Open history ± double-blinded food challenges

Prevalence and natural history

In the young child, fruits and fruit juices may cause minor gastro-intestinal reactions. Fruits and vegetables, particularly tomatoes, strawberry and citrus fruits also cause perioral rashes, usually eczema and urticaria, in early childhood. If one includes these minor non-IgE mediated skin and gastro-intestinal symptoms, fruits and fruit juices were the foods most commonly causing adverse reactions in one study with a cumulative incidence of 12% in children less than three years of age, with 58% of reactions occurring to orange juice, tomato juice and apple (Bock 1987). Other studies suggest a cumulative incidence of adverse reactions to citrus fruits of around 3–5% in the first three years of life (Saarinen and Kajosaari 1980, Kajosaari 1982), with a higher rate of 13% if the definition is based on history rather than food challenge (Saarinen and Kajosaari 1980).

In contrast to the minor reactions of early childhood, IgE-mediated reactions occur later, so that up to 75% of IgE reactions to fruits and vegetables have their onset after two years of age (Crespo *et al.* 1995). In an allergy clinic based study, fruits and vegetables were responsible for the vast majority of IgE-mediated food allergy presenting after the age of ten years (Kivity *et al.* 1994). Many of these later childhood reactions occur in a subgroup of children with pollen sensitisation resulting in cross-reactivity to a range of fruits. This food allergy presents as a contact allergy with oral symptoms, known as the oral allergy syndrome, and occurs mainly with raw fruit or vegetables. The prevalence of allergy to different fruits and vegetables varies with the type and amount of pollen present, which determines the cross-reacting fruits and vegetables. Thus in Scandinavia, with its high levels of birch pollen, there is a high prevalence of apple allergy (Foucard 1991), whereas in Japan where there is more Japanese cedar, the allergy is mainly to melon and kiwi (Arai *et al.* 1998).

Regarding the natural history of adverse reactions to fruits and vegetables, clinical reactivity is short-lived in those children with onset in early childhood. In one study, tolerance to fruits and fruit juices was achieved after a mean of 15 months (range 3–34 months, median 13 months) (Bock 1987). The natural history data for the later onset predominantly IgE-mediated reactions are not well defined, but are certainly of longer duration.

10.3.8 Chocolate

Chocolate

A 2.2–6.6% self-reporting of reactions to chocolate are reported in two questionnaire surveys (Niestijl Jansen *et al.* 1994, Young *et al.* 1994). In an American population-based birth cohort study of 480 children followed up to three years old, 1.7% complained of adverse reactions to chocolate, but none was confirmed on food challenge (Bock 1987), and chocolate is rarely a cause of positive food challenge in allergy clinics (Bock *et al.* 1988, Crespo *et al.* 1995). It is likely that the majority of the reported reactions are to other components in the chocolate, for example cows' milk and nuts.

10.3.9 Food additives

The commonest food additives thought to cause adverse reactions are tartrazine (E102), sunset yellow (E110), annatto, aspartame, benzoic acid and sulphites (Fuglsang *et al.* 1993). Key epidemiological studies are shown in Table 10.9. Adverse reactions to food additives can occur at any age. A UK study showed a higher reporting of adverse reactions to food additives in the first ten years of life, and more often occurring in females (Young *et al.* 1987). The mechanism of the reaction is often unknown, and IgE-mediated reactions are rare. Questionnaire-based studies give a high 6.6–7.4% prevalence of self-reported adverse reactions to food additives in the general population. However, when food challenges are used to make the diagnosis, the prevalence falls to about 0.23%. One study shows the risk to be greatest in the atopic population, with no reactions observed in non-atopic individuals (Fuglsang *et al.* 1994). Virtually all reactions are minor and limited to the skin (worsening of eczema/urticaria) with serious systemic reactions rarely reported. Regarding the natural history, there are no good epidemiological data.

10.3.10 Interpreting data on the natural history of food allergy

Food allergies can be divided into groups with similar natural histories (Figure 10.2). Cohort studies have been very successful in delineating the natural history of allergies to foods such as cows' milk and egg because they are almost completely outgrown within a few years. For longer lived allergies, such as fish, shellfish, peanut and tree nuts, the natural history is less clear because of the difficulties in interpreting the available data. This is illustrated by results from an interview survey investigating the prevalence of peanut allergy (Emmett *et al.* 1999). The data (Figure 10.3) suggest that more males are affected in childhood whereas in adulthood peanut allergy is more prevalent in females. There are a number of possible explanations for these results. Firstly, peanut allergy may be outgrown at an earlier age in males. Secondly, peanut allergy may be acquired later in females. Thirdly, there may be a combination of both of the above. Fourthly, the data may be explained by a cohort effect: the adult generation surveyed may have a lower inherent risk of peanut allergy than the childhood generation such that if the survey was repeated in 15 years' time, a greater adult prevalence would be seen as these children become adults. And lastly, the results may have also been subjected to information bias as a house-to-house survey will primarily sample adult females who may not be aware that adult male co-habitants have an allergy to peanuts thereby reducing the apparent prevalence in adult males. Even if these issues can be resolved, we are left only knowing the prevalence of peanut allergy which is a summation of existing cases, new cases and remissions. In order to overcome these problems, prospective cohort studies are needed where the point prevalence is established longitudinally together with the incidence and remission rates.

Table 10.9 Epidemiology of adverse food reactions to food additives – key studies

Author, date	Type of study	Country	No. of subjects	Point prevalence	Definition
Young *et al*. 1987	Population based, cross-sectional	UK	18 582	7.4% at > 6 months old	History
Fuglsang *et al*. 1994	Population based, cross-sectional	Denmark	4274	6.6% at 5–16y	History
Fuglsang *et al*. 1994	Non-atopic population, cross-sectional	Denmark	4274	0% at 5–16y	Open challenge
Fuglsang *et al*. 1994	Atopic population, cross-sectional	Denmark	4274	9.8% at 5–16y	Open challenge
Young *et al*. 1987	Population based, cross-sectional	UK	649	0.23% at > 4 y	History and DBPCFC

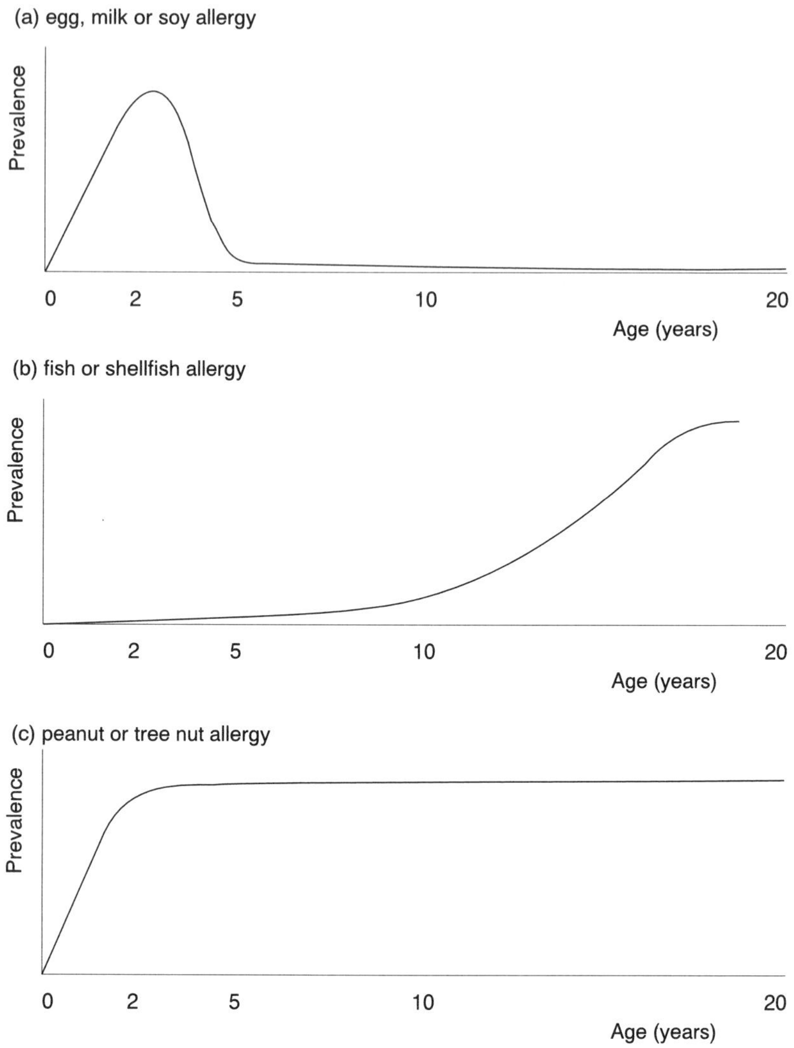

Figure 10.2 Natural histories of different food allergies.

10.4 Geographical variations

Data concerning the incidence of adverse food reactions from different countries may shed some light on factors that might be important in the development of adverse food reactions. These factors include genetic, cultural, dietary and other environmental differences. Unfortunately all the cohort studies are from Europe, Australia and the USA, with no comparable data from other countries. However, there are case series from these other countries that allow comparisons to be made between foods that are important in causing adverse reactions in different countries.

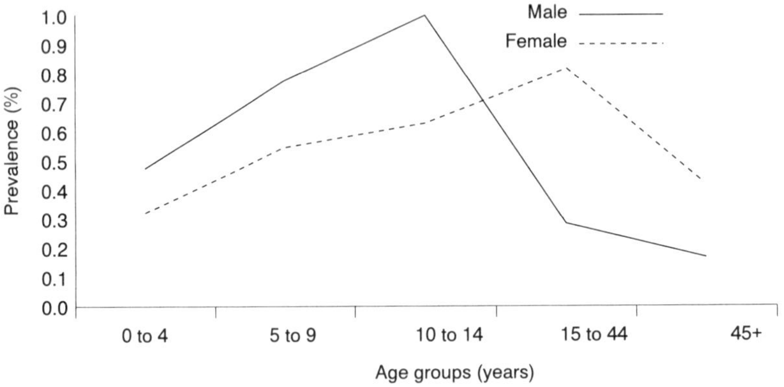

Figure 10.3 Prevalence of peanut allergy in males and females (Emmett *et al.* 1999).

10.4.1 Common food allergies

Table 10.10 compares clinical reactions to foods, and Table 10.11 compares skinprick/specific IgE reactions to foods, between allergy clinic populations from different countries. As such they deal with a selected population and some studies involve small numbers. They show that cows' milk and egg are among the 2–3 commonest foods causing allergy in most countries. Peanut, fish, soy, wheat and shellfish are among the next most common groups of foods causing allergy, although significant variations occur between countries. Thus, for example, shellfish allergy appears to be more common in countries such as the Philippines, Thailand and Singapore where it is a part of the staple diet from early infancy, than in many other countries where it is consumed later and less commonly. In contrast, clinical peanut allergy which is a big problem in Western countries appears to be less common in most Asian countries, and also in Spain (Crespo *et al.* 1995). Thus in Japan it is very rare (Hill *et al.* 1997), and in one study from Singapore no cases of nut allergy were seen in 124 consecutive admissions with anaphylaxis (Goh *et al.* 1999). Positive skinprick test to peanut appears to be less common in many Asian countries, especially in Japan, although it accounted for 10% of positive skin tests in a Singapore allergy clinic (Shek and Lee 1999).

Table 10.10 also shows variations in fruit and vegetable allergy with higher rates in Mediterranean countries. Thus peach allergy is common in a Spanish study (Crespo *et al.* 1995, Cuesta-Herranz *et al.* 1998) and accounts for 44% of 112 positive food challenges in patients between 10 and 48 years of age in an Israel allergy clinic (Kivity *et al.* 1994). In both countries, peach consumption is high.

Coeliac disease also has geographical variations. Although formerly thought to be a disease associated with north-west Europeans, including the countries of their migration, chiefly the USA, an equivalent prevalence is being reported in other European countries (Greco *et al.* 1989). It appears to be uncommon

Table 10.10 Foods causing clinical allergic reactions in allergy clinics from different countries. (Data presented as percentage of total reactions.)

Spain Crespo *et al.* 1995	France Rance *et al.* 1999	USA Bock *et al.* 1988	Japan Hill *et al.* 1997	Philippines Hill *et al.* 1997
608 foodstuffs, SPT & sp IgE & history, children	813 foodstuffs, SPT/sp IgE & FC, children	481 DBPCFCs in 323 children	79 FCs, mean age 5.2y	46 FCs, mean age 2.0y
Egg 20.1%	Egg 35.7%	Egg 33.1%	Egg 59.4%	Cows' milk 32.6%
Fish 17.8%	Peanut 23.6%	Peanut 18.7%	Cows' milk 22.8%	Shellfish 26.1%
Cows' milk 14.3%	Cows' milk 8.3%	Cows' milk 18.1%	Wheat 12.6%	Egg 21.7%
Lentils 5.9%	Mustard 6%	Soy 6.7%	Soy 5.1%	Fish 10.9%
Peaches 5.1%	Cod 4.3%	Nuts 4.8%		Wheat 4.3%
Peanut 3.9%	Hazelnut 1.8%	Shellfish 4.4%		Peanut 2.2%
Chick peas 3.9%	Kiwi 1.5%	Fish 4.0%		Soy 2.2%
Crustaceans 3.8%	Wheat 1.5%	Wheat 3.3%		

FC: food challenge

Table 10.11 Foods causing sensitisation (positive skin prick/specific IgE tests) in paediatric allergy clinics in Australia and Asia. (Modified from Hill *et al*. 1997. Data presented as percentage of total reactions.)

Australia	Hong Kong	China	Thailand	Japan	Philippines	Singapore	Taiwan
Egg 37%	Cows' milk 41%	Egg 64%	Shellfish 22%	Egg 48%	Fish 25%	Shellfish 36%	Egg 30%
Cows' milk 23%	Egg 29%	Cows' milk 14%	Peanut 21%	Cows' milk 18%	Shellfish 22%	Cows' milk 16%	Cows' milk 21%
Peanut 22%	Wheat 10%	Peanut 9%	Soy 16%	Wheat 12%	Wheat 15%	Egg 16%	Peanut 19%
Sesame 5%	Fish 10%	Soy 9%	Rice 16%	Fish 9%	Egg 12%	Wheat 14%	Soy 12%
Cashew 4%	Soy 9%	Wheat 4%	Egg 15%	Soy 8%	Cows' milk 12%	Peanut 10%	Shellfish 9%
Hazelnut 2%	Peanut 4%		Cows' milk 10%	Rice 4%	Peanut 8%	Soy 8%	Wheat 7%
Walnut 2%				Peanut 1%	Soy 6%		

elsewhere, and seems not to occur in Afro-Caribbeans or Chinese (Ferguson 1999).

10.4.2 Novel and uncommon food allergies

There are a number of foods that are eaten in geographically or culturally quite specific populations and adverse food reactions are limited to these groups. However, with diversification of cultures and diets across the globe, particularly in developed countries, adverse reactions to these foods may be seen in many other countries. A good example is sesame seed, to which allergy in Western countries was rarely reported (Rance *et al.* 1999). However, there are reports of an increasing number of cases of sesame seed allergy in France coincident with the increase in Middle Eastern food and fast food bread (Kolopp-Sarda *et al.* 1997). Sesame seed often causes severe clinical allergy hence its importance. In France sesame seed was responsible for 0.6% of IgE-mediated food allergies seen in recent years in an allergy clinic population (Rance *et al.* 1999).

Table 10.12 makes the point that uncommon food allergens are important causes of food allergy in specific countries. In an Israel allergy clinic population, sunflower seed was responsible for 22.3% of 112 positive food challenges in subjects between 10 and 48 years of age (Kivity *et al.* 1994). In Singapore, out of 124 consecutive admissions with anaphylaxis, the commonest cause was bird's nest soup (Goh *et al.* 1999), a food not implicated in allergy elsewhere in the world. In Japan rice appears to be a relatively common cause of allergy causing atopic eczema, although more severe acute reactions to rice are rare (Ikezawa *et al.* 1992). Rice is also a common cause of food allergy in Thailand (Hill *et al.* 1997). Adverse reaction to buckwheat is a common problem in Japan. In a population of 92,680 schoolchildren in Japan, the incidence of adverse reaction to buckwheat on questionnaire was 0.22% (Takahashi *et al.* 1998). The risk of anaphylaxis to buckwheat was higher than for egg and milk. In Hong Kong, royal jelly consumption is common with 31.3% of 461 hospital employees surveyed in one study consuming it (Leung *et al.* 1997); 7.4% of the subjects had a positive skinprick test to pure royal jelly, 0.6% had a history of clinical allergy, and nearly all employees with a positive skinprick test also had other atopic features. Pineapple allergy is responsible for a reported 23.5% of food allergy in Indonesia (Hill *et al.* 1997).

Table 10.12 Some uncommon foods causing allergy in specific countries

France	Japan	Hong	Singa- Kong	Spain pore	Israel	Indonesia	Poland
Lentil Mustard Snail	Rice Buck-wheat	Royal jelly	Bird's nest soup	Lentil	Sun-flower seed	Pineapple Chicken	Beef

In Spain, with its Mediterranean diet, lentils are the legumes most commonly implicated in allergy (Pascual *et al.* 1999), and a common cause of anaphylaxis. Chick peas, peas, and green beans are also not uncommonly seen as causes of food allergy in Spain, with lentils, chick peas and beans forming 4.9% of the protein diet in Spain. In France, lentil is responsible for 0.8% of clinical food allergy seen in an allergy clinic, having not been seen in previous years (Rance *et al.* 1999). Lentils do not appear in the list of the 481 positive food challenges in a large USA allergy clinic based study (Bock *et al.* 1988). Mustard is an important cause of allergy in France, accounting for 6% of all food allergies seen in an allergy clinic (Rance *et al.* 1999), but does not appear to be a problem in other countries.

Beef allergy is a problem in Poland where it ranks among the six foods most likely to cause allergy (Czaja-Bulsa and Bachorska 1998), and is also a problem in China (Hill *et al.* 1997). Chicken is a common reported cause of allergy in Indonesia (Hill *et al.* 1997). In other countries, meats are often reported by subjects as causes of adverse reactions, but rarely confirmed by food challenge. Thus, in UK and Dutch questionnaire studies, 1.6–2.7% of respondents from random populations reported adverse reactions to meat and meat products (Young *et al.* 1994, Niestijl Jansen *et al.* 1994). However, this high figure has not been substantiated by studies involving food challenges in which adverse reactions to meats were found to be rare (Bock 1987, Crespo *et al.* 1995). Snail allergy is reported only from France, Spain and Portugal where it is eaten as a delicacy (de la Cuesta *et al.* 1989). Many of the allergic reactions are severe, involving respiratory compromise.

10.5 Cross-reactions between foods

Cross-reactivity is due to a reaction to identical or similar protein allergens that occur in more than one food, or in a food and an inhalant pollen. This is different from associated reactivity where two or more food allergens may be seen to be associated epidemiologically. A good example of the latter is the high rate of association between egg and peanut allergy although the allergens are not related. Establishing a cross-reaction requires the demonstration of at least a positive correlation between the magnitude of specific IgE to both foods, and RAST inhibition studies are needed for confirmation. Cross-reactivity is seen at an immunological level when a subject is sensitised to both foods on the basis of positive skinprick or specific IgE testing to both foods. However, often only a smaller proportion will demonstrate clinical cross-reactivity, that is a reaction to both foods on clinical exposure.

Table 10.13 lists the common cross-reactions involving foods. For fish and legumes, there are good data regarding cross-reactivity at immunological (skin prick/specific IgE) and clinical levels. One study demonstrated 73% immunological cross-reactivity for ten different fish species, but only 28% clinical cross-reactivity to two or more of the same ten species (Bernhisel-Broadbent *et al.* 1992). In the case of legumes, the same authors demonstrated immunological

Table 10.13 Common cross-reactions involving foods

Index food or pollen	Cows' milk	Chicken egg	Cod	Shrimp	Peanut	Latex
Cross-reacting foods	Soy 11–35% clinical cross-reaction Sheep and goat milk 50–75% clinical cross-reaction	Duck, geese & turkey egg	Other fish 28% clinical cross-reaction 73% skinprick/ IgE cross-reaction	Crustaceans, molluscs	Other legumes e.g. soya bean, garden pea, chick pea, lentil, guar, liquorice, carob, gum arabic and other beans 5–60% clinical cross-reaction 75% skinprick/ IgE cross-reaction	Fruits and vegetables e.g. banana, pear, avocado, chestnut, papaya, potato, tomato
References	Juntunen and Ali-Yrkko 1983 Bardare *et al*. 1988 Zeiger *et al*. 1999	Langeland 1983	Bernhisel-Broadbent *et al*. 1992	Musmand *et al*. 1993	Bernhisel-Broadbent and Sampson 1989 Crespo *et al*. 1995	Lavaud *et al*. 1992 Beezold *et al*. 1996

Table 10.13 continued

Index food or pollen	Birch pollen	Ragweed pollen	Mugwort pollen	Grass pollen
Cross-reacting foods	Fruits and vegetables e.g. apple, celery, carrot, potato, kiwi, hazelnut, cherry 5–60% clinical cross-reaction 10–75% skinprick/IgE cross-reaction	Fruits and vegetables e.g. watermelon, melon, cucumber, banana	Legumes (see peanut) Also celery, carrot, nuts, mustard	Tomato, potato, green pea, peanut, watermelon, melon, apple, orange, kiwi
References	Dreborg and Foucard 1983 Foucard 1991 Caballero *et al.* 1994	Ortolani *et al.* 1998	Caballero and Martin-Esteban 1998	Caballero and Martin-Esteban 1998

cross-reactivity between legumes in 49 out of 69 patients (71%) with atopic eczema, but only 2 out of 41 patients (5%) evaluated showed clinical cross-reactivity (Berhisel-Broadbent and Sampson 1989). By way of contrast, another study showed considerable clinical cross-reactivity between the legumes in the context of acute reactions (Crespo *et al.* 1995). In the latter study, out of 67 patients seen in an allergy clinic, 43 (64%) showed clinical allergic reactions to more than one legume (mainly lentil, peanut, chick pea, pea and bean).

In Scandanavian countries there is a high prevalence of birch pollen sensitisation, reaching up to 10–15% in teenagers and young adults (Eriksson 1978). Between 30% and 75% report clinical reactivity to fruits and vegetables, occurring chiefly as the oral allergy syndrome in adolescents and adults, with apple being the food most commonly implicated (Dreborg and Foucard 1983, Pastorella *et al.* 1995). Clinical cross-reactivity can be confirmed in around 60–75% of birch pollen allergic patients with immunological cross-reactivity to foods (Foucard 1991). These data are from patient history and food challenges done at home, and a lower reaction rate is likely with more rigorous food challenge procedures (Caballero *et al.* 1994).

10.6 Occupational food allergy

There are a number of subjects who are at increased risk of developing food allergy related to occupational exposure, virtually all mediated by an IgE reaction. The most common and best studied foods are listed below.

10.6.1 Shellfish

A number of shellfish can cause occupational asthma and rhinoconjunctivitis primarily from inhalation of particles during food processing. The reactions have been demonstrated to occur with shrimp, crab and oyster handlers. Workers affected include those involved in seafood processing, cooks and fishermen (Malo and Cartier 1993). Up to 10–40% of workers exhibit respiratory symptoms, and in studies where skin testing has been done up to 60% are found to be positive, with a close correlation between skin test reactivity and clinical reactivity (Orford and Wilson 1985, Cartier *et al.* 1986).

10.6.2 Flour

Baker's asthma is due to sensitisation to cereal proteins. The majority of cases are reported to wheat, rye and barley, and it has been one of the most common occupational diseases in the UK (Block *et al.* 1984). One study found 7–9% of bakers to be affected (Thiel and Ulmer 1980), and there may be a long latent period of up to 10–15 years before symptoms occur. Again, atopic individuals appear to be at increased risk (Prichard *et al.* 1985).

10.6.3 Legumes and seeds

Various legumes and seeds can cause occupational asthma. The most common and best studied is the coffee bean, caused by the green (unroasted) coffee bean, which produces a positive skin test in up to 82%, and a clinical reaction in up to 10%, of exposed workers (Osterman *et al.* 1985). Cotton seed, linseed and tea dust can all cause occupational asthma (Cartier and Malo 1990).

10.6.4 Eggs

Egg proteins may cause occupational asthma in 5–20% of workers involved in egg processing, and atopy again appears to be a risk factor (Bernstein *et al.* 1987, Smith 1990).

10.7 Risk factors for the development of adverse food reactions

For all the risk factors that may be involved in the development of adverse food reactions, much of the epidemiology has concentrated on asthma, eczema or total allergy. Few studies look specifically at risk factors for adverse food reactions. This section concentrates on the literature concerning adverse food reactions, and more specifically IgE-mediated food allergy.

10.7.1 Associated morbidity

Markers of atopy as a whole are associated with an increased risk of developing adverse food reactions. Thus asthma, eczema and rhinitis are increased in children with food allergy compared to the general population (Zeiger and Heller 1995, Hide *et al.* 1996). The strongest association is between eczema and food allergy, and the risk appears to be greatest in infancy and in those with moderate to severe eczema (Burks *et al.* 1998, Sampson 1996). The literature appears to be best for peanut allergy. One study found that in peanut-allergic children atopy in some other form was present in up to 96% of subjects (Ewan 1996). In the Isle of Wight birth cohort study half of the children with peanut allergy had asthma and two-thirds had eczema, considerably higher than the rates in the cohort as a whole (Tariq *et al.* 1996).

10.7.2 Immunological markers

The role of cord blood total IgE as a marker for the development of food allergy is not clear. Studies do not consistently show a positive association (Dean 1997, Kjellmann *et al.* 1988, Kulig *et al.* 1999). Furthermore, in the recent German multicentre allergy study where an association between cord blood total IgE and sensitisation to foods at one year of age was found, the authors comment on the poor predictive performance of cord blood IgE (Kulig *et al.* 1999). This study puzzlingly also showed that an elevated cord blood total IgE was a significant protective factor for early-onset atopic eczema (Edenharter *et al.* 1998). Thus, cord blood total IgE is an unhelpful marker in predicting the development of food allergy and in planning appropriate prevention strategies.

Prenatal sensitisation with antigen-specific IgE has been reported but seems to be uncommon, and limited to cows' milk (Businco *et al.* 1983, Host *et al.* 1992). It is therefore unlikely to play a role in the vast majority of food allergy. None of the large birth cohort studies have demonstrated any specific IgE to foods in cord blood, including cows' milk, even in children who subsequently developed clinical or immunological sensitisation (Kjellmann *et al.* 1988, Hide *et al.* 1996, Kulig *et al.* 1999). It is therefore not surprising to find that dietary intervention in pregnancy has shown no benefit in modifying the natural history of IgE-mediated food allergy. Food sensitisation as measured by allergen-specific T-cell responses have been demonstrated in cord and fetal blood (Jones *et al.* 1998). However, none of these responses have been assessed to be risk factors in the development of food allergy. It is unclear whether these responses are derived from a memory T-cell population or from a naive population.

10.7.3 Genetic factors

Family history of atopy

If either parent has a history of an allergic disease then siblings are at increased risk of developing allergic disease, which includes eczema, asthma, allergic rhinitis and food allergy (Zeiger and Heller 1995). The risk is greater if either parent is atopic, and increases if both parents are atopic. In children with cows' milk allergy, a family history of atopy in first-degree relatives has been found in 23–80% of cases (Goldman 1963, Ventura 1988, Host 1990). Findings from a Danish study looking at skin reactions to foods are presented in Table 10.14, confirming the association of food allergy and family history of atopy (Kjellman 1983).

A family history of food allergy in a first-degree relative increases the risk of food allergy approximately fourfold in other family members (Dean 1997). In families with at least two food allergic individuals, the same food is frequently implicated. The best-studied food is peanut whereby if one sibling has peanut allergy then the risk of another sibling having peanut allergy is 7%. This represents a tenfold increased risk compared with the general population in whom the risk is 0.6% (Tariq *et al.* 1996). However, there is a lack of good literature looking at the risk for other foods and in general there are no studies, such as twin studies, that separate the role of genetic and environmental factors in the development of food allergy.

Associated food allergy

In view of the association between food allergy and atopy, it is not surprising to find an individual with food allergy having allergy to one or more foods, even in the absence of cross-reactivity. This associated reactivity is seen for egg and peanuts (Dean 1997), and is best studied with cows' milk. Children with cows' milk allergy have a 7–58% chance of developing egg allergy, 0–16% chance of developing cereal allergy and a 0–35% chance of developing allergy to orange (Host 1995). Associated reactivity in cows' milk allergic subjects is also seen to tomato and banana (Host 1995). The Melbourne Atopy Cohort Study which prospectively followed an atopic birth cohort of 100 children who developed cows' milk allergy found evidence of allergy to one or more other foods in up to 75% of subjects in the five years of follow-up (Hill *et al.* 1999). These associated food allergies are more likely to reflect the underlying atopic tendency of the individual rather than constituting an additional and specific risk factor for the development of food allergy.

HLA studies

A number of different HLA genotypes have been shown to be associated with different types of food allergy. The data are best for peanut allergy and coeliac disease (Howell *et al.* 1998, Howdle and Blair 1992).

Gene linkage studies

Linkage of genes on a number of chromosomes with various atopic markers such as asthma and eczema has been demonstrated (Cookson *et al.* 1989, Nickel *et al.* 1999). However, no specific markers for food allergy have been established and none of the candidate genes has shown clinical application in terms of predicting or preventing the development of food allergy.

Environmental factors

The following have been shown to be statistically significant risk factors, after multivariate analysis, for the development of food allergy in the first 1–7 years of life: a history of both parents smoking, male gender, 4-month IgE $\geq$ 1 SD, and nasal eosinophils at one year of age (Arshad *et al.* 1992, Zeiger and Heller 1995). Many other factors have been shown to be associated with the development of aeroallergen sensitisation, asthma or total allergy, but not specifially food allergy. These include prematurity, season of birth, pets, maternal asthma, either parent smoking, defects in lymphocyte regulation, immune response genes, specific and early infections with viruses, and exposure to products of pollution (Dean 1997, Lucas *et al.* 1990, Zeiger and Heller 1995).

Table 10.14 History of food allergy before seven years of age in comparison with a family history of atopy. (Adapted from Kjellmann 1983.)

Atopic parents	Total % of children with food allergy
2	58
1	29
0	13

10.8 Intervention strategies aimed at preventing adverse food reactions

Most of the work in this area has been directed at preventing allergic sensitisation (primary prevention), rather than the prevention or suppression of clinical disease once sensitisation has occurred (secondary and tertiary prevention respectively). Up to now, no therapy has been shown to be of value in secondary or tertiary prevention of adverse food reactions. Furthermore, whilst some studies show that pharmacological intervention may alter the incidence and natural history of asthma, there are no comparable data regarding adverse food reactions (Bustos *et al.* 1995, Warner 1997). This section therefore concentrates on the dietary intervention studies set up with the aim of preventing or reducing the occurrence of adverse food reactions. Some of the studies look at children with a high risk of atopy (usually defined as those children with at least one first-degree relative with documented atopic disease), others at unselected children from the general population. Most do not focus specifically on reaction to food as an endpoint, and this section concentrates on the few studies that use immunological (skinprick or specific IgE) testing to foods or clinical reactivity using food challenge as endpoints for the preventative strategies. Some of the well-conducted intervention studies using eczema as an end-point are also mentioned, in view of the association of eczema with adverse food reactions in the early years of life.

10.8.1 Maternal intervention

Intervention in pregnancy
The potential for *in utero* sensitisation to food allergens via the placenta or swallowing of amniotic fluid has led to a number of investigators restricting possible antigens in the maternal diet during pregnancy.

Intervention during lactation
The potential for sensitisation during lactation also exists as small amounts of food allergens have been found in breast milk. Beta-lactoglobulin is found in the breast milk of 95% of mothers consuming cows' milk during lactation (Host *et al.* 1988). Peanut and other proteins have also been found in breast milk (Bock 1982). Such observations have led investigators to assess the allergy prevention effects of a restricted maternal diet during lactation.

There are other studies where the maternal dietary intervention occurred during pregnancy and lactation.

Using the endpoints of clinical food reactions and immunological sensitisation, there is no evidence from the available studies (Table 10.15) to suggest that dietary restriction in pregnancy reduces the risk of the infant developing adverse food reactions, in either normal or high-risk subjects. One study has suggested a possible protective effect of maternal peanut avoidance in pregnancy and lactation in an atopic population (Hourihane and Kilburn 1997). This study was

Table 10.15 Prospective randomised studies assessing the effect of maternal dietary intervention in pregnancy and lactation on the development of adverse food reactions

Study (reference)	No. and type of subjects	Maternal diet	Follow-up period (yrs)	Definition of adverse food reaction	Outcome
Kjellman *et al*. 1988	210, atopic population, birth cohort	No egg, no cows' milk in 3rd trimester vs. no restriction	1.5	SPTs to egg and cows' milk in 1/3 of subjects	No difference
Lilja *et al*. 1991	162, atopic population, birth cohort	Reduced egg and cows' milk in 3rd trimester and 1st 2 months of lactation vs. no restriction	1.5	SPTs and sp IgE to ovalbumin, ovomucoid and beta-lactoglobulin Total IgE	No difference
Falth-Magnusson and Kjellman 1992	197, atopic population, birth cohort	No egg, no cows' milk in 3rd trimester vs. no restriction	5	SPTs and sp IgE	No difference
Sigurs *et al*. 1992 Hattevig *et al*. 1996	115, atopic population, birth cohort, groups assigned by hospital rather than true randomisation	No egg, no cows' milk, no fish in 1st 3 months of lactation vs. no restrictions	10	History of intolerance to cows' milk/egg SPT/sp IgE to egg, cows' milk, hazelnut, peanut, fish, wheat Eczema	No difference in history and SPT/sp IgE Less eczema in prophylaxis group at 4y but not at 10y

retrospective and uncontrolled, but suggested that in an atopic population the consumption of peanuts by mothers during pregnancy and lactation was associated with an earlier onset of peanut allergy in the children. There was no difference in the cumulative incidence of peanut allergy, and timing of immunological sensitisation to peanut was not assessed. An alternative explanation of the data is that the children of mothers consuming peanuts during pregnancy and lactation had the opportunity to consume peanuts earlier in life than those whose mothers did not eat peanuts. Furthermore, the findings of this study are not supported by a study based on the Isle of Wight birth cohort (Tariq *et al.* 1996). This study showed no effect of reduced/no maternal nut ingestion in pregnancy on the development of immunological or clinical reaction to nuts in a non-randomised population followed up until four years of age.

Using eczema as the endpoint, which of course may or may not be associated with adverse food reactions, a number of studies in atopic populations using maternal dietary restriction during lactation alone (Chandra *et al.* 1989, Lovegrove *et al.* 1994, Hattevig *et al.* 1996) or during the last trimester of pregnancy and lactation (Chandra *et al.* 1986) have shown a reduction in eczema. The protective effect lasts for between 18 months and four years, with no effect being seen on ten-year follow-up (Hattevig *et al.* 1996). Not all the studies are randomised, and two of the studies have an unusually high prevalence of eczema in the control (no dietary restriction) population (Chandra *et al.* 1986, Lovegrove *et al.* 1994).

In conclusion there is no consistent evidence to support maternal pregnancy dietary restriction in an attempt to reduce the risk of adverse food reactions. This is not surprising given the studies showing an absence of specific IgE to foods in cord blood (Kjellmann *et al.* 1988, Hide *et al.* 1996, Kulig *et al.* 1999). Although in infants from an atopic population the risk of eczema in the short to medium term may be reduced by dietary restriction during lactation, there is no long-term benefit and no association with reduced adverse food reactions.

10.8.2 Infant intervention

Breast feeding vs. cows' milk vs. other milks

There are large variations between the studies comparing the different milks, namely breast milk, soya, hypoallergenic formulae (partially or extensively hydrolysed cows' milk), and cows' milk formulae given to the infant and the development of allergy. Many of the studies have looked for effects of the type of infant milk feeding on the development of allergic respiratory or skin disease, rather than on food or immunological (skinprick/specific IgE) reactions. A number of the early studies attempting to look at the impact of the infant milk formula on the risk of developing adverse food reactions showed a marginal reduction in skin test reactions and clinical adverse reactions to cows' milk (Hamburger 1984, Host *et al.* 1988, Saarinen and Kajosaari 1995). However, many were not randomised or prospective in their design. The more recent studies which attempt to look at the impact on adverse food reactions are randomised (Table 10.16) and use food challenge and skinprick/specific IgE endpoints, and occasionally eczema, as markers for adverse food reactions.

Of the studies listed in Table 10.16 only one shows a reduced specific IgE and clinical reactivity to milk in the intervention group with a partially hydrolysed formula, an effect that disappeared after 6 months of age (Vandenplas *et al.* 1995). The one other study that did suggest a reduction in reactions to cows' milk in breast/hydrolysate-fed babies versus unmodified cows' milk using open food challenge is flawed by the intervention group having a different year of recruitment for the control (cows' milk) group (Halken *et al.* 1993). The other studies do not consistently support any link between the type of infant milk feed and the development of adverse food reactions if skinprick/specific IgE and food challenge criteria are applied.

A number of studies in Table 10.16 use eczema, which in the early years may be associated with food intolerance, as the endpoint. The studies consistently show a protective effect of breast milk or cows' milk based hydrolysates versus unmodified cows' milk based formula on the development of eczema in the first 12–48 months of life in an atopic population (Chandra and Hamed 1991, Mallet and Henocq 1992, Vandenplas *et al.* 1995, Oldaeus *et al.* 1997). Only one small study looking at a normal population suggests a benefit of breast milk over cows' milk in reducing the risk of eczema, but with only short-term follow-up (Lucas *et al.* 1990). The data do not consistently support any benefit of breast feeding over a hydrolysed formula, nor do they favour an extensively hydroysed formula over a partially hydrolysed one. Soy-based formulas confer no protective benefit, and no evidence supports the use of goat or sheep milk which immunologically cross-react with cows' milk (Miskelly *et al.* 1988, Chandra *et al.* 1989, Chandra and Hamed 1991).

In conclusion, international studies do not suggest the view that different infant formulae or prolonged breast feeding reduce the risk of IgE mediated milk or other food allergies. However, there is a consistent view from a number of studies, particularly in regard to the atopic population, that breast milk and milk hydrolysates do reduce the risk of developing eczema in early childhood, an effect that disappears after 4–5 years of age.

Introduction of solids

There are no good prospective randomised studies looking specifically at the effect of delaying the introduction of solids on the risk of adverse food reactions. Prospective, non-randomised studies from a normal population (Fergusson *et al.* 1990) and an atopic population (Kajosaari 1991) have shown that delayed introduction of solid foods for 4–6 months reduced the risk of eczema. The study using a normal population showed a risk of chronic/recurrent eczema 2.9 times greater in those infants fed four or more solid foods before the age of four months compared with infants receiving no solid foods before four months of age. This difference was maintained until ten years of age (Fergusson *et al.*

Table 10.16 Prospective, randomised trials assessing the effect of infant milk feeding on the development of adverse food reactions

Study (reference)	No and type of subjects	Infant milk	Follow-up period (yrs)	Definition of adverse food reaction	Outcome
Lucas *et al*. 1990	75, preterm, population based	Breast milk vs. preterm cows' milk for 1.5 months	1.5	Eczema	Reduced eczema in breast-fed group up to 18 months
Chandra and Hamed 1991	288, atopic population, birth cohort	Whey hydrolysate vs. cows' milk vs. soya vs. breast fed > 4 months	1.5	Sp IgE/SPT to cows' milk and soya Eczema	No difference in sp IgE/SPTs Increased eczema in cows' milk and soya groups
Schmitz *et al*. 1992	256, population based, birth cohort	Cows' milk vs. partially hydrolysed casein for first few days in breast-fed babies	1	History Sp IgE to cows' milk Eczema	No difference
Mallet and Henocq 1992	165, atopic population, birth cohort	Casein hydrolysate vs. cows' milk for 4 months	4	Sp IgE to cows' milk Eczema	No difference in sp IgE Reduced eczema up to 4y
Vandenplas *et al*. 1995	58, atopic population, birth cohort	Partially hydrolysed whey vs. cows' milk for 6 months	7	Open FC Sp IgE to cows' milk at 6 months Eczema	At 6 months 33% of control group had intolerance to cows' milk vs. 4% in intervention group (p=0.006). No difference at 1y. Reduced cows' milk IgE in intervention group at 6 months Reduced eczema in intervention group up to 1y

Oldaeus *et al*. 1997	155, atopic population, birth cohort	Partially hydrolysed (PH) vs. extensively hydrolysed (EH) vs. cows' milk (CM) vs. breast fed (BF) for 9 months	1.5	Open or DBPC food challenges in 20% SPT & sp IgE to egg & cows' milk Eczema	No difference in positive FC Increased SPT to egg in PH group at 9 months but not at 18 months Increased eczema in CM and PH up to 9 months and in CM at 18 months, compared with BF & EH groups
De Jong *et al*. 1998	1533, population based, birth cohort	Cows' milk vs. protein-free formula for first 3 days in breast-fed babies	2	History Sp IgE to egg & cows' milk Eczema	No difference

1990). In the study of an atopic population, eczema and a history of food allergy were reduced at the age of one year in the group fed solids after six months of age compared with those with solids introduced at three months. No food challenges or skinprick/IgE testing were performed in the first year, but at five years there was no difference between skin testing to fish, milk and wheat, history of food allergy and eczema between the two groups (Kajosaari 1991). A randomised, population-based study in Finland showed no difference in the cumulative incidence of fish and citrus allergy at three years old between children with fish introduced early or late (after one year old) into the diet, although the children with earlier introduction reacted earlier in life (Saarinen and Kajosaari 1980). Similar observations have been reported with coeliac disease. The later introduction of gluten into the infant diet has altered the age of onset and type of clinical presentation of coeliac disease in countries such as the UK and Scandanavia, but does not seem ultimately to stop the development of the disease, a view supported by the increase in serological population screening studies (Logan 1992, Ascher 1996, Hallert 1998).

Various guidelines exist in the UK recommending delayed introduction of solids in infants at increased risk of atopy, and in the same at-risk group delaying the ingestion of peanut products until after three years of age (Committee on Toxicity of Chemicals in Food 1998). On the basis of the studies presented, these guidelines do not appear to be evidence-based. Furthermore, the observation that 88% of egg reactions and 80% of peanut reactions occur after the first known exposure (Ford and Taylor 1982, Hourihane and Kilburn 1997) suggests that allergen avoidance is not straightforward and sensitisation may occur earlier in life and by other means, such as food contamination or inhaled sensitisation (Witteman 1995).

In conclusion, the evidence to date suggests that delaying the introduction of a solid food will perhaps postpone rather than prevent the development of clinical food allergy. There are no data suggesting that immunological (skin test or specific IgE) reactivity is affected. Thus, at the age of five years no difference in sensitisation to foods between those with solids introduced early or late into the diet can be found (Kajosaari 1991). These observations are probably not surprising as a delay in the age at which clinical reactivity develops may simply reflect the timing of the food being introduced into the diet, thereby giving the individual the first opportunity to clinically react to the food. Although there is some evidence that delaying the introduction of solids to 4–6 months reduces the risk of eczema in the medium term, the data come from non-randomised studies, and thus have to be interpreted with caution.

Combined maternal and infant measures

Two of the best trials in the field of dietary avoidance involve combined maternal and infant interventions (Zeiger and Heller 1995, Hide *et al.* 1996). Both are prospective and randomised with assessments by physicians blinded to the randomisation group in an atopic population. Both used skin test/specific IgE and food challenge criteria as endpoints for adverse food reactions, as well as other

Table 10.17 Prospective, randomised trials assessing the effect of mixed maternal and infant dietary measures on the development of adverse food reactions

Study (reference)	No. and type of subjects	Maternal and infant diet	Follow-up period (yrs)	Definition of AFR	Outcome
Zeiger *et al*. 1989 Zeiger and Heller 1995	288, atopic population, birth cohort, randomised, physician blinded	Maternal egg, cows' milk and peanut avoidance in 3rd trimester and lactation + infant breast or casein hydrolysate (6 months), cows' milk and solids delayed > 6 months (later for some solids) vs. American Academy of Pediatrics guidelines	7	DBPCFC (50% of subjects) Sp IgE/SPT to cows' milk, egg, wheat, corn, soy, peanut, cod, chicken/beef	Reduced food intolerance at 1y and reduced cows' milk IgE/SPT at 1y and 2y in intervention group Reduced eczema at 1y in intervention group
Arshad *et al*. 1992 Hide *et al*. 1996	120, atopic population, birth cohort, randomised, physician blinded	Maternal egg, cows' milk, fish and nuts exclusion during lactation + infant breast +/− soy hydrolysate, solids delayed >11 months vs. no restrictions	4	Open challenge SPT to cows' milk, egg, wheat, fish, peanut Eczema	Differences in prevalence of cows' milk/egg intolerance and food SPTs did not reach statistical significance Reduced eczema until 4y in intervention group

atopic diseases including eczema (Table 10.17). One study involved maternal dietary restriction in pregnancy and lactation, infant breast or casein hydrolysate feeding for six months, and delayed introduction of solids until at least six months into the infant diet (Zeiger and Heller 1995). There was a reduction in adverse food reactions in the intervention group at one year of age using a combination of clinical history and DBPCFC for diagnosis. These differences were almost entirely due to cows' milk allergy. The effect had disappeared by two years of age. The intervention group also showed a significant reduction in cows' milk specific IgE and cows' milk skinprick test at one and two years of age, but not thereafter. There were equal numbers sensitised to peanut at all ages including seven years when this was the commonest positive food allergen. Eczema was reduced at one year in the intervention group but not thereafter. The second study from the Isle of Wight cohort (Hide *et al.* 1996) involved maternal food avoidance during breast feeding and infant cows' milk avoidance until nine months with breast or soya hydrolysate used until then and egg introduced as the first solid at 11 months. There were reduced numbers of subjects with positive food challenges and food skin tests, mostly at one year old but never reaching statistical significance. Eczema was reduced until the four-year follow-up.

The conclusions from combined maternal and infant dietary exclusions are of a reduction in cows' milk allergy until 1–2 years of age, and a reduction in eczema in the first 1–4 years of life. As the natural history of cows' milk allergy is one of natural resolution by the age of two years in the vast majority, it is not surprising that the effect of dietary avoidance on food allergy disappears by two years of age. These studies on combined exclusion diets show no long-term benefit in preventing egg, peanut and other persistent food allergies.

10.9 Conclusions

The measured incidence and prevalence of adverse food reactions in a population depend largely on the precise definition and diagnostic criteria. The gold standard for diagnosing adverse food reactions is the DBPCFC but this is not suited to large epidemiological studies for practical reasons. In such studies, specific IgE alone will measure allergic sensitisation rather than clinical allergy and overestimate the true incidence and prevalence of food allergy. In such large population studies, the combination of a specific clinical history for food allergy together with specific IgE determination or SPT provides a more accurate measure of food allergy in the population.

The measurement of incidence and prevalence in birth cohort studies provides the most reliable epidemiological data; the bias inherent in other study designs is considerably reduced and problems of interpretation due to cohort effects are diminished. Such prospective studies also allow accurate description of the natural history of adverse food reactions.

Regrettably, there are few such studies and those that have been performed have been in European or other developed countries. Nevertheless, comparative data on the relative importance of adverse food reactions in different populations can be derived from case series that rank the relative importance of different food allergies seen in specialist allergy clinics. Important observations emerge from such comparative data. Firstly, egg and milk allergies are the most common food allergies world-wide. Secondly, certain food allergies that are common in Western countries, such as peanut allergy, may be uncommon in Asian countries such as Japan. Thirdly, certain food allergies that are never seen or are extremely rare in Western countries are important causes of allergy in other countries. This forces us to rethink our concept of 'common' and 'uncommon' allergenic foods. Different food allergens are clinically important in different countries: mustard allergy in France; sunflower seed allergy in Israel; lentil allergy in Spain; royal jelly allergy in Hong Kong; and bird's nest allergy in Singapore. Fourthly, it emerges that foods described as 'hypoallergenic' may be important allergens in countries outside Western Europe and North America. Thus beef allergy is important in Poland, chicken allergy is important in Hong Kong and rice allergy is a significant problem in Japan.

The erosion in the distinction between common and uncommon food allergens and between allergenic and hypoallergenic foods has important implications for the food industry. The 'globalisation' of eating habits and introduction of new foods into different cultures, e.g. kiwi fruit, sesame and mango, is likely to lead to changes in the pattern of food allergies seen across the world, with new, previously rare, allergies occurring with increasing frequency in different countries. Additionally, there are considerations to be taken into account in the development of novel foods, especially when derived from genetically modified organisms (GMOs). Existing proposals to evaluate the safety of GMOs with respect to food allergy depend on whether the transgenic protein is derived from a common or uncommon allergenic source. In the light of the above, such a distinction seems artificial at best and is likely to be misleading. We are not justified in dismissing the risk posed by a transgenic protein derived from an 'uncommon' food allergen.

It is an interesting fact that the frequency of adverse food reactions in a population is clearly related to its presence in the local diet and its early introduction in infancy, as demonstrated by the previously cited examples of common and uncommon food allergens. It is therefore surprising that dietary intervention aimed at delaying the introduction of a food into a child's diet fails to reduce the prevalence of food allergies. Although dietary intervention during pregnancy and lactation is clearly able to reduce infantile eczema and delay its onset, there is no convincing evidence that it significantly prevents the development of food allergies, with the exception of cows' milk. Importantly, several studies fail to demonstrate the presence of specific IgE to most food allergens in cord blood. This argues strongly against allergic sensitisation being completed *in utero* and suggests that the transplacental passage of allergen may not play an important role. Cows' milk allergy is the exception since specific IgE to beta lactoglobulin and other cows' milk proteins have been detected in cord blood. This perhaps explains why combined dietary intervention during pregnancy and lactation prevents the development of cows' milk allergy. However, as far as other food allergens are concerned, maternal dietary exclusion and delayed introduction into the infant's diet merely delays the manifestation of food allergy but does not appear to inhibit the development of allergic sensitisation and subsequent clinical allergy.

In summary, important epidemiological work needs to be done with respect to food allergy. An international effort, similar to the International Study of Asthma and Allergies in Childhood (ISAAC), would be a useful approach. This would ideally employ concurrent birth cohort studies in different parts of the globe. Such studies would yield important data on the world-wide prevalence of different food allergies and provide important clues to the pathogenesis of food allergy with the discovery of novel interventional strategies.

10.10 References

AMERICAN ACADEMY OF ALLERGY AND IMMUNOLOGY/NIAID (1984) *Adverse reactions to foods* (eds Anderson J A, Sogn D D), pp1–6, NIH Publication 84-2442, Washington.

ANDERSON C M (1992) The evolution of a successful treatment for coeliac disease. In Marsh M N ed, *Coeliac Disease* 1–16. Blackwell, Oxford.

ARAI Y, OGAWA C, OHTOMO M, ITO K (1998) Food and food additives hypersensitivity in adult asthmatics. II. Oral allergy syndrome in adult asthmatics with or without Japanese cedar hay fever. *Arerugi – Japanese Journal of Allergology* 47(8): 715–19.

ARSHAD S H, STEVENS M, HIDE D W (1993) The effect of genetic and environmental factors on the prevalence of allergic disorders at the age of two years. *Clin Exp Allergy* 23: 504–11.

ARSHAD S H, MATTHEWS S, GANT C, HIDE D W (1992) Effect of allergen avoidance on development of allergic disorders in infancy. *Lancet* 339: 1493–7.

ASCHER H (1996) The role of quality and quantity of gluten containing cereals in the epidemiology of coeliac disease. In: *Coeliac Disease Proceedings of the seventh international symposium on coeliac disease.* Tampere, Finland, 15–22.

BARDARE M, MAGNOLFI C, ZANI G (1988) Soy sensitivity: personal observation of 71 children with food intolerance. *Allerg Immunol* 20: 63–6.

BEEZOLD D H, SUSSMANN G L, LISS G M, CHANG N S (1996) Latex allergy can induce clinical reactions to specific foods. *Clin Exp Allergy* 26: 996–9.

BERNHISEL-BROADBENT J, SAMPSON H A (1989) Cross-reactivity in the legume botanical family in children with food hypersensitivity. *J Allergy Clin Immunol* 83: 435–40.

BERNHISEL-BROADBENT J, SCANLON S M, SAMPSON H A (1992) Fish hypersensitivity. I. *In vitro* and oral challenge results in fish-allergic patients. *J Allergy Clin Immunol* 89: 730–7.

BERNSTEIN D I, SMITH A B, MOLLER D R (1987) Clinical and immunological studies among egg-processing workers with occupational asthma. *J Allergy Clin Immunol* 80: 791–7.

BJORNSSON E, JANSON C, PLASCHKE P, SJOBERG O (1996) Prevalence of sensitisation to food allergens in adult Swedes. *Annals of Allergy, Asthma and Immunology* 77: 327–32.

BLOCK G, TSE KS, KIJEK K, CHAN H, CHAN-YEUNG M (1984) Baker's asthma: studies of the cross-antigenicity between different cereal grains. *Clin Allergy* 14: 177–85.

BOCK SA (1982) The natural history of food sensitivity. *J Allergy Clin Immunol* 69: 173–7.

BOCK SA (1987) Prospective appraisal of complaints of adverse reactions to foods in children during the first 3 years of life. *Pediatrics* 79: 683–8.

BOCK SA (1992) Incidence of severe food reactions in Colorado (abstract). *J Allergy Clin Immunol* 89: 192.

BOCK SA ATKIN FM (1989) The natural history of peanut allergy. *J Allergy Clin Immunol* 83: 900–4.

BOCK SA SAMPSON HA, ATKINS RM *et al.* (1988) Double-blinded, placebo-controlled food challenge as an office procedure: a manual. *J Allergy Clin Imunol* 82: 986–97.

BRUINJZEEL-KOOMEN CAFM, ORTOLAI C, AAS K *et al.* (1995) Adverse reactions to foods. Positions paper. *Allergy* 50: 623–36.

BRUNO G, GIAMPIETA PG, DEL GEURCIO MJ, GALLIA P, GIOVANNINI L, LOVATI C, PAOLUCCI P, QUAGLIO L, ZORATTI E, BUSINCO L (1997) Soy allergy is not common in atopic children: a multicentre study. *Ped Allergy Immunol* 8: 190–3.

BURKS AW, JAMES JM, HIEGEL A, SAMPSON H (1998) Atopic dermatitis and food hypersensitivity reactions. *J Pediatr* 132: 132–6.

BURR ML, MERRETT TG (1983) Food intolerance: a community survey. *Br J Nutr* 49: 217–19.

BUSINCO L, LUCCENTI P, GIAMPIETRO PG (1995) Allergenicity of goat's milk in children with cows' milk allergy. *Allergologie* 18: 412–13.

BUSINCO L, MARCHETTI F, PELLEGRINI G, PERLINI R (1983) Predictive value of cord blood IgE levels in 'at risk' newborn babies and influence of type of feeding. *Clin Allergy* 13(6): 503–8.

BUSTOS GJ, BUSTOS D, ROMERO O (1995) Prevention of asthma with ketotifen in infants with atopic dermatitis. *Clin Exp Allergy* 25: 568–73.

CABALLERO T, MARTIN-ESTEBAN M (1998) Association between pollen hypersensitivity and edible vegetable allergy: a review. *J Invest Allergology & Clin Immunol* 8(1): 6–16.

CABALLERO T, MARTIN-ESTEBAN M, GARCIA-ARA C, PASCUAL C, OJEDA A (1994) Relationship between pollinosis and fruit or vegetable sensitization. *Pediatr Allergy Immunol* 5: 218–22.

CANTANI A, LUCENTI P (1997) Natural history of soy allergy and/or intolerance in children, and clinical use of soy-protein formulas. *Pediatr Allergy Immunol* 8: 59–74.

CARROCCIO A, CAVATAIO F, IACONO G (1999) Cross-reactivity between milk proteins of different animals. *Clin Exp Allergy* 29: 1014–16.

CARTIER A, MALO J-L (1990) Occupational asthma due to tea dust. *Thorax* 45: 203–6.

CARTIER A, MALO J-L, FOREST F (1986) Occupational asthma in snow crab processing workers. *J Allergy Clin Immunol* 78: 344–8.

CASTILLO R, CARRILO T, BLANCO C, QUIRALTE J, CUEVAS M (1994) Shellfish hypersensitivity: clinical and immunological characteristics. *Allergol et Immunopathol* 22: 83–7.

CATASSI C, RATSCH IM, FABIANI E (1994) Coeliac disease in the year 2000: exploring the iceberg. *Lancet* 343: 200–3.

CHANDRA R K, HAMED A (1991) Cumulative incidence of atopic disorders in high risk infants fed whey hydrolysate, soy, and conventional cow milk formulas. *Ann Allergy* 67: 129–32.

CHANDRA R K, PURI S, HAMED A (1989) Influence of maternal diet during lactation and use of formula feeds on development of atopic eczema in high risk infants. *Br Med J* 299: 228–30.

CHANDRA R K, PURI S, SURAIYA C, CHEEMA P S (1986) Influence of maternal food antigen avoidance during pregnancy and lactation on the incidence of atopic eczema in infants. *Clin Allergy* 16: 563–9.

COMMITTEE ON TOXICITY OF CHEMICALS IN FOOD, CONSUMER PRODUCTS AND THE ENVIRONMENT (1998) *Peanut allergy*. London: Department of Health.

COOKSON W O C M, FAUX J A, SHARP P A, HOPKIN J M (1989) Linkage between immunoglobulin E responses underlying asthma and rhinitis and chromosome 11q. *Lancet* 1: 1292–4.

CRESPO J F, PASCUAL C, BURKS A W, HELM R M ESTEBAN M M (1995) Frequency of food allergy in a pediatric population from Spain. *Pediatr Allergy Immunol* 6: 39–43.

CUESTA-HERRANZ J, LAZARO M, DE LAS HERAS M, LLUCH M, CUESTA C (1998) Peach allergy pattern: experience in 70 patients. *Allergy* 53: 78–82.

CZAJA-BULSA G, BACHORSKA J (1998) Food allergy in children with pollinosis in the Western sea coast region. *Polski Merkuriusz Lekarski* 5(30): 338–40.

DEAN T (1997) Prevalence of allergic disorders in early childhood. *Ped Allergy Immunol* 8(suppl 10): 27–31.

DE JONG M H, SCHARP-VAN DEN LINDEN V T M, AALBERSE R C, OOSTING J, TIJSSEN J G P, DE GROOT C J (1998) Randomised controlled trial of brief neonatal exposure to cows' milk on the development of atopy. *Arch Dis Child* 79: 126–30.

DE LA CUESTA C G, GARCIA B E, CORDOBA H, DIEGUEZ I, OEHLING A (1989) Food allergy to Helix terrestre (snail). *Allergologia et Immunopathologia* 17(6): 337–9.

DE MARTINO M, NOVEMBRE E, GALLI L, DE MARCO A, BOTARELLI P, MARANO E, VIETUCCI A (1990) Allergy to different fish species in cod-allergic children: *in vivo* and *in vitro* studies. *J Allergy Clin Immunol* 86: 909–14.

DREBORG A, FOUCARD T (1983) Allergy to apple, carrot and potato in children with birch pollen allergy. *Allergy* 38: 167–72.

EDENHARTER G, BERGMANN RL, BERGMANN KE (1998) Cord blood IgE as a risk factor and predictor for atopic diseases. *Clin Exp Allergy* 28: 671–9.

EMMETT SE, ANGUS FJ, FRY JS, LEE PN (1999) Perceived prevalence of peanut allergy in Great Britain and its association with other atopic conditions and with peanut allergy in other household members. *Allergy* 54: 380–5.

ERIKSSON NE (1978) Food sensitivity reported by pateints with asthma and hay fever. *Allergy* 33: 299–309.

EWAN PW (1996) Clinical study of peanut and nut allergy in 62 consecutive patients: new features and associations. *Br Med J* 312: 1074–8.

FALTH-MAGNUSSON K, KJELLMANN NIM (1992) Allergy prevention by maternal elimination diet during pregnancy: a 5-year follow-up study. *J Allergy Clin Immunol* 89: 709–13.

FERGUSON A (1999) The coeliac iceberg. *CME Journal Gastroenterology Hepatology and Nutrition* 2: 52–6.

FERGUSSON DM, HORWOOD J, SHANNON FT (1990) Early solid feeding and recurrent childhood eczema: a 10-year longitudinal study. *Pediatrics* 86: 541–6.

FORD RPK, TAYLOR B (1982) Natural history of egg hypersensitivity. *Arch Dis Child* 57: 649–52.

FOUCARD T (1991) Allergy and allergy-like symptoms in 1050 medical students. *Allergy* 46: 20–6.

FRANK L, MARIAN A, VISER M, *et al.* (1999) Exposure to peanuts *in utero* and in infancy and the development of sensitisation to peanut allergies in young children. *Ped Allergy Immunol* 10: 27–32.

FUGLSANG G, MADSEN C, SAVAL P, OSTERBALLE O (1993) Prevalence of intolerance to food additives among Danish children. *Pediatr Allergy Immunol* 4: 123–9.

FUGLSANG G, MADSEN C, HALKEN S *et al.* (1994) Adverse reactions to food additives in children with atopic symptoms. *Allergy* 49: 31–7.

GERRARD JW, MACKENZIE JWA, GOLUBOFF N *et al.* (1973) Cows' milk allergy. Prevalence and manifestations in an unselected series of newborn. *Acta Paed Scand* Suppl 234: 1–21.

GIAMPIETRO PG, RAGNO V, DANIELE S *et al.* (1992) Soy hypersensitivity in children with food allergy. *Annals of allergy* 69: 143–6.

GOH DL, LAU YN, CHEW FT, SHEK LP, LEE BW (1999) Pattern of food-induced anaphylaxis in children of an Asian community. *Allergy* 54(1): 84–6.

GOLDING J, FOX DES, LACK G (1998) Prevalence and natural history of peanut allergy in children in the UK. *J Allergy Clin Immunol* 101: S103.

GOLDMAN AS (1963) Milk allergy. *Pediatrics* 32: 425–43.

GOLDMAN AS, ANDERSON DW, SELLERS WA *et al.* (1963) Milk allergy I. Oral challenge with milk and isolated milk proteins in allergic children. *Pediatrics* 32: 425–43.

GRECO I, TOZZI AE, MAYER M (1989) Unchanging clinical picture of coeliac disease presentation in Campania, Italy. *Eur J Pediatr* 148: 610–13.

GUINNEPAIN M, ELOIT C, RAFFARD M, RASSEMONT R, LAURENT J (1996) Exercise-induced anaphylaxis: useful screening of food sensitisation. *Ann Allergy Asthma Immunol* 77: 491–6.

HALKEN S, HOST A, HANSEN LG, OSTERBALLE O (1993) Preventative effect of feeding high-risk infants a casein hydrolysate formula or an ultrafiltrated whey hydrolysate formula. A prospective, randomised comparative clinical study. *Pediatr Allergy Immunol* 4: 173–81.

HALPERN SR, SELLARS WA, JOHNSON RB *et al.* (1973) Development of childhood allergy in infants fed breast, soy or cow milk. *J Allergy Clin Immunol* 51: 139–51.

HALLERT C (1998) The epidemiology of coeliac disease: a continuous enigma. In: *The Changing Features of Coeliac Disease*. The Finnish Coeliac Society. Tampere, Finland. Eds Lohiniemi S, Collin P, Maki M.

HAMBURGER RN (1984) Diagnosis of food allergies and intolerances in the study of prophylaxis and control groups in infants. *Ann Allergy* 53: 673–7.

HANSEN TK, BINDSLEVJENSEN C, SKOV PS, POULSEN LK (1997) Codfish allergy in adults: IgE cross-reactivity among fish species. *Ann Allergy Asthma Immunol* 78: 187–94.

HATTEVIG G, SIGURS N, KJELLMANN B (1996) Maternal food antigen avoidance during lactation and allergy during the first 10 years of age. *J Allerg Clin Immunol* 97(3): 241.

HELBLING A, HAYDEL A, McCANT SML, MUSMAND JT, EL-DAHR J, LEHRER SB (1999) Fish allergy: is cross reactivity among fish species relevant? Double blind, placebo controlled food challenge studies of fish allergic adults. *Ann Allergy Asthma Immunol* 83: 517–23.

HERMANN M-E, DANNEMANN A, GRUTERS A, WAHN U (1996) Prospective study on the atopy preventative effect of maternal avoidance of milk and eggs during pregnancy and lactation. *Eur J Pediatr* 155: 770–4.

HIDE DW, MATHEWS S, TARIQ S, ARSHAD SH (1996) Allergen avoidance in infancy and allergy at 4 years of age. *Allergy* 51: 89–93.

HILL DJ, BALL G, HOSKINGS CS (1988) Clinical manifestations of cows' milk allergy in childhood I. Associations with *in vitro* cellular immune responses. *Clin Allergy* 18: 469–79.

HILL DJ, HOSKING CS, ZHIE CY, LEUNG R, BARATWIDJAJA K, IIKURA Y, IYNGKARAN N, GONZALEZ-ANDAYA A, WAH LB, HSIEH KH (1997) The frequency of food allergy in Australia and Asia. *Environmental Toxicol Pharm* 4: 101–10.

HILL DJ, HOSKINGS CS, HEINE RG (1999) Clinal spectrum of food allergy in children in Australia and South-east Asia: identification and targets for treatment. *Ann Med* 31: 272–81.

HINS H, BIRD G, FISCHER P, MAHY N, JEWELL D (1999) Coeliac disease in primary care:a case finding study. *Br Med J* 318: 164–7.

HOST A (1990) A prospective study of cow milk allergy. *Allergy* 45: 587–96.

HOST A (1995) Adverse reactions to food: epidmiology and risk factors. *Ped Allergy Immunol* 6(8): 20–8.

HOST A, HALKEN S (1990) A prospective study of cow milk allergy in Danish infants during the first 3 years of life. Clinical course in relation to clinical and immunological type of hypersensitivity reaction. *Allergy* 45: 587–96.

HOST A, HALKEN S (1998) Epidemiology and prevention of cows' milk allergy. *Allergy* 53(Suppl 46): 111–13.

HOST A, HALKEN S, JACOBSON H P *et al.* (1997) The natural course of cows' milk protein allergy/intolerance. *J Allergy Clin Immunol* 99: S490.

HOST A, HUSBY S, OSTERBALLE O (1988) A prospective study of cows' milk allergy in exclusively breast-fed infants. *Acta Paediatr Scand* 77: 663–70.

HOST A, HUSBY S, GJESING B, LARSEN J N, LOWENSTEIN H (1992) Prospective estimation of IgG, IgG subclass and IgE antibodies to dietary proteins in infants with cow milk allergy. *Allergy* 47: 218–29.

HOURIHANE J O'B, KILBURN S A, DEAN P, WARNER J O (1997) Clinical characteristics of peanut allergy. *Clin Exp Allergy* 27: 634–9.

HOURIHANE J O'B, ROBERTS S A, WARNER J O (1998) Resolution of peanut allergy: case controlled study. *Br Med J* 316: 1271–5.

HOWDLE P D, BLAIR G E (1996) Molecular biology and coeliac disease. *Gut* 33: 573–5.

HOWELL W M, TURNER S J, HOURIHANE J O, DEAN T P, WARNER J O (1998) HLA class II DRB1, DQB1 and DPB1 genotypic associations with peanut allergy: evidence from a family-based and case-control study. *Clin Exp Allergy* 28(2): 156–62.

HUTTLY S R, BARROS F C, VICTORA C G, BERIA J U, VAUGHAN J P (1990) Do mothers overestimate breast feeding duration? An example of recall bias from a study in Southern Brazil. *Am J Epimediol* 132: 572–5.

IKEZAWA Z, MIYAWAKA K, KOMATSU H, SUGA C, SUZUKI Y (1992) A probable involvement of rice allergy in severe type atopic dermatitis in Japan. *Acta Dermato-Veneorologica.* Suppl. 176: 103–7.

JAKOBSON I, LINDBERG T (1979) A prospective study of cows' milk protein intolerance in Swedish infants. *Acta Paed Scand* 68: 853–9.

JONES C A, KILBURN S A, WARNER J A, WARNER J O (1998) Intrauterine environment and fetal allergic sensitization. *Clin Exp Allergy* 28: 655–9.

JUNTUNEN K, ALI-YRKKO S (1983) Goats' milk for children allergic to cows' milk. *Kiel Milchwirt Forschungsber* 35: 439–40.

KAJOSAARI M (1982) Food allergy in Finnish children aged 1 to 6 years. *Acta Paediatr Scand* 71: 815–19.

KAJOSAARI M (1991) Prospective 5-year follow-up study of children with six months exclusive breast feeding and solid food elimination. *Adv Exp Med Biol* 310: 453–8.

KANNY G, DE HAUTECLOCQUE C, MONERET-VAUTRIN D A (1996) Sesame seed and sesame seed oil contain masked allergens of growing importance. *Allergy* 51: 952–7.

KEMP S F, LOCKEY R F, WOLF B L, *et al.* (1995) Anaphylaxis: a review of 266 cases. *Arch Intern Med* 155: 1749–54.

KIVITY S, DUNNER K, MARIAN Y (1994) The pattern of food hypersensitivity in patients with onset after 10 years of age. *Clin Exp Allergy* 24: 19–22.

KJELLMAN N-IM (1983) Development and prediction of atopic allergy in childhood. In: Bostrom H, Ljungstedt N, eds. *Skandia International Symposia. Theoretical and clinical aspects of allergic diseases*. Stockholm: Almqvist & Wicksell, 57–73.

KJELLMAN N-IM, BJORKSTEN B, HATTEVIG G, FALTH-MAGNUSSON K (1988) Natural history of food allergy. *Ann Allergy* 61(2): 83–7.

KOLOPP-SARDA M N, MONERET-VAUTRIN D A, GOBERT B (1997) Specific humoral response in 12 cases of food sensitization to sesame seed. *Clin Exp Allergy* 27: 1285–91.

KUITUNEN A, VISAKORPI J K, SAVILAHTI E *et al.* (1975) Malabsorption syndrome with cows' milk intolerance. Clinical findings and course in 54 cases. *Arch Dis Child* 50: 351–6.

KULIG M, BERGMANN R, KLETTKE U, WAHN U (1999) Natural course of sensitization to food and inhalant allergens during the first 6 years of life. *J Allergy Clin Immunol* 103: 1173–9.

LANGELAND T (1983) A clinical and immunological study of allergy to hen's egg white. IV. Occurrence of proteins cross-reacting with allergens in hen's egg white as studied in egg white from turkey, duck, goose, seagull and in hen's egg yolk and hen and chicken sera and flesh. *Allergy* 38: 399–412.

LAVAUD F, COSSART C, REITER V (1992) Latex allergy in patients with allergy to fruit. *Lancet* 339: 22–4.

LEUNG R, HO A, CHAN J, CHOY D, LAI C K (1997) Royal Jelly consumption and hypersensitivity in the community. *Clin Exp Allergy* 27(3): 333–6.

LILJA G, DANNAEUS A, FOUCARD T, GRAFF-LONNEVIG V, JOHANSSON S G O, OMAN H (1991) Effects of maternal diet during late pregnancy and lactation on the development of IgE and egg- and milk-specific IgE and IgG antibodies in infants. *Clin Exp Allergy* 21: 195–202.

LOGAN R F A (1992) Problems and pitfalls in epidemiological studies of coeliac disease. In: Auricchio S and Visakorpi J K, eds. *Common Food Intolerances 1: Epidemiology of Coeliac Disease*. Basel: Karger, 14–22.

LOVEGROVE J A, HAMPTON S M, MORGAN J B (1994) The immunological and long-term atopic outcome of infants born to women following a milk-free diet during pregnancy and lactation: a pilot study. *Br J Nutr* 71: 223–38.

LOZA C, BROSTOFF J (1995) Peanut allergy. *Clin Exp Allergy* 25: 493–502.

LUCAS A, BROOKE O G, COLE T J, MORLEY R, BAMFORD J T M (1990) Early diet of preterm infants and development of allergic or atopic disease: randomised prospective study. *Br Med J* 300: 837–40.

LUCAS E W (1979) Food uses of peanut proteins. *J Am Oil Chem Soc* 56: 425–30.

MAGNOLFI C, ZANI G, LACAVA L *et al.* (1996) Soy allergy in atopic children. *Ann Allergy Asthma Immunol* 77: 197–201.

MALLET E, HENOCQ A (1992) Long-term prevention of allergic diseases by using protein hydrolysate formula in at-risk infants. *J Pedatr* 121: S95–S100.

MALO J-L, CARTIER A (1993) Occupational reactions in the seafood industry. *Clin Rev Allergy* 11: 223–40.

MATRICARDI P M, ROSMINI F, RIONDINO S, FORTINI M, FERRIGNO M, RAPICETTA M,

BONINI S (2000) Exposure to foodborne and orofecal microbes versus airborne viruses in relation to atopy and allergic asthma: epidemiological study. *Br Med J* 320: 412–17.

MISKELLY FG, BURR ML, VAUGHAN-WILLIAMS E, FEHILY AM, BUTLAND BK, MERRETT TG (1988) Infant feeding and allergy. *Arch Dis Child* 63: 388–93.

MUNOZ-FURLONG A, SICHERER SH, BURKS AW, SAMPSON HA (1989) Prevalence of peanut and tree nut allergy in the United States. *J Allergy Clin Immunol* 101: S103.

MUSMAND JJ, DAUL CB, LEHRER SB (1993) Crustacea allergy. *Clin Exp Allergy* 23: 722–32.

NAVARRO C, MARQUEZ M, HERNANDO L *et al.* (1993) Epidemic asthma in Cartagena, Spain, and its association with soyabean sensitivity. *Epidemiology* 34: 76–9.

NICKEL R, BEYER K, HUANG SK, BARNES KC, WAHN U (1999) Genetic markers of atopy in infancy: results from the German Multicenter Allergy Study. *Clin Exp Allergy* 29 (suppl 4): 23–5.

NICKEL R, KULIG M, FORSTER J, BERGMANN R, BAUER CP, LAU S, WAHN U (1997) Sensitization to hen's egg at the age of twelve months is predictive for allergic sensitization to common indoor and outdoor allergens at the age of three years. *J Allergy Clin Immunol* 99(5): 613–17.

NIESTIJL JANSEN JJ, KARDINAAL AFM, HUIJBERS G *et al.* (1994) Prevalence of food allergy and intolerance in the adult Dutch population. *J Allergy Clin Immunol* 93: 446–56.

OLDAEUS G, ANJOU K, BJORKSTEIN B, MORAN JR, KJELMANN N-IM (1997) Extensively and partially hydrolysed infant formulas for allergy prophylaxis. *Arch Dis Child* 77: 4–10.

ORFORD RR, WILSON JT (1985) Epidemiological and immunological studies in processors of the King crab. *Am J Industr Med* 7: 155–69.

ORTOLANI C, ISPANO M, ANSALONI R, ROTONDO F, PASTORELLA EA (1998) Diagnostic problems due to cross-reactions in food allergy. *Allergy* 53(suppl 46): 58–61.

OSTERMAN K, JOHANSSON SGO, ZETTERSTROM O (1985) Diagnostic tests in allergy to green coffee. *Allergy* 40: 336–43.

PAGANUS A, JUNTUNEN-BACKMAN K, SAVILAHTI E (1992) Followup of nutritional status and dietary survey in children with cows' milk allergy. *Acta Paed* 81: 518–21.

PASCUAL CY, FERNANDEZ-CRESPO J, SANCHEZ S, PADIAL A, DIAZ-PENA JM, MARTIN-ESTEBAN M (1999) Allergy to lentils in Mediterranean pediatric patients. *J Allergy Clin Immunol* 103: 154–8.

PASTORELLA EA, INCORVAIA C, ORTOLANI C (1995) The mouth and pharynx. *Allergy* 50 (suppl 20): 53–5.

PERKKIO M, SAVILAHTI E, KUITUNEN P (1981) Morpholometric and immunohistochemical study of jejunal biopsies from children with intestinal soy allergy. *Eur J Pediatr* 137: 63–9.

POWELL G K (1978) Milk and soy-induced enterocolitis of infancy: clinical features and standardizations of challenge. *J Pediatr* 93: 553–60.

PRICHARD M G, RYAN H, WALSH B J, MUSK A W (1985) Skin test and RAST responses to wheat and common allergens and respiratory disease in bakers. *Clin Allergy* 15: 203–10.

RANCE F, KANNY G, DUTAU G *et al.* (1999) Food hypersensitivity in children: Clinical aspects and distribution of allergens. *Pediatr Allergy Immunol* 10: 33–8.

RATNER B, UNTRACT S (1952) Egg allergy in children. *Am J Dis Child* 83: 309–16.

SAARINEN U M, KAJOSAARI M (1980) Does dietary elimination in infancy prevent or only postpone a food allergy? *Lancet* 166–7.

SAARINEN U M, KAJOSAARI M (1995) Breastfeeding as prophylaxis against atopic disease: prospective follow-up study until 17 years old. *Lancet* 346: 1065–9.

SACKETT D L (1979) Bias in analytical research. *J Chronic Diseases* 32: 51–63.

SAMPSON H A (1988) The role of food hypersensitivity and mediator release in atopic dermatitis. *J Allergy Clin Immunol* 81: 635–45.

SAMPSON H A (1990) Peanut anaphylaxis. *J Allergy Clin Immunol* 86: 1–3.

SAMPSON H A (1996) Epidemiology of food allergy (review). *Pediatr Allergy Immunol* 7: S42–S50.

SAMPSON H A (1998) Legumes, eggs and milk. *Allergy* 53 (suppl 46): 38–43.

SAMPSON H A, McCASKILL C C (1985) Food hypersensitivity and atopic dermatitis: evaluation of 113 patients. *J Pediatr* 107: 669–75.

SAMPSON H A, SCANLON S M (1989) The natural history of food hypersensitivity in children with atopic dermatitis. *J Pediatr* 115: 23–7.

SAMPSON H A, MENDELSON L, ROSEN J P (1992) Fatal and near-fatal anaphylactic reactions to food in children and adolescents. *N Eng J Med* 327: 380–4.

SCHMITZ J, DIGEON B, CHASTANG C, STROBEL S (1992) Effects of brief early exposure to partially hydrolysed and whole cow milk proteins. *J Pediatr* 121(5Pt2): S85–9.

SCHRANDER J J P, BOGART J H P, FORGET P P *et al.* (1993) Cows' milk protein intolerance in infants under 1 year of age: a prospective epidemiological study. *Eur J Pediatr* 152: 640–4.

SHEK L P C, LEE B W (1999) Food allergy in children – The Singapore story. *Asian Pacific J Allergy Immunol* 17: 203–6.

SIGURS N, HATTEVIG G, KJELLMANN B (1992) Maternal avoidance of eggs, cows' milk and fish during lactation: Effect on allergic manifestations, skinprick tests and specific IgE antibodies in children at age 4 years. *Pediatr* 89: 735–9.

SMITH A B (1990) Evaluation of occupational asthma from airborne egg protein exposure in multiple settings. *Chest* 98: 398–402.

TAKAHASHI Y, ICHIKAWA S, AIHARA Y, YOKOTA S (1998) Buckwheat allergy in 90,000 schoolchildren in Yokohama. *Aerugi-Japanese Journal of Allergology* 47(1): 26–33.

TARIQ S M, STEVENS M, MATTHEWS S, RIDOUT S, TWISELTON R, HIDE D W (1996) Cohort study of peanut and tree nut sensitisation by age of 4 years. *Br Med J* 313: 514–17.

THIEL H, ULMER W T (1980) Baker's asthma: development and possibility for treatment. *Chest* 78: 400–5.

VANDENPLAS Y, HAUSER B, VAN DEN BORRE C, DAB I (1995) The long-term effect of a partial whey hydrolysate formula on the prophylaxis of atopic disease. *Eur J Pediatr* 154: 488–94.

VARJONEN E, KALIMO K, LAMMINTAUSTA K, TERHO P (1992) Prevalence of atopic disorders among adolescents in Turku, Finland. *Allergy* 47: 243–8.

VENTURA A (1988) Cows' milk allergy in the first year of life. *Acta Paediatr Scand* 388 (suppl.): 1–14.

WARNER J O (1997) Early treatment of the atopic child. *Pediatr Allergy Immunol* 8 (10 suppl): 46–8.

WITTEMAN A M (1995) Food allergens in house dust. *Int Arch Allergy Immunol* 107: 566–8.

YOUNG E, PATEL S, STONEHAM M, RONA R, WILKINSON J D (1987) The prevalence of reaction to food additives in a survey population. *J Royal College Phys* 21(4): 241–9.

YOUNG E, STONEHAM M D, PETRUCKEVITCH A, BARTON J, RONA R (1994) A population study of food intolerance. *Lancet* 343: 1127–30.

YUNGINGER J W, SWEENEY K G, STURNER W Q, *et al.* (1988) Fatal food-induced anaphylaxis. *JAMA* 260: 1450–2.

ZEIGER R S, HELLER S (1995) The development and prediction of atopy in high-risk children: Follow-up at age seven years in a prospective randomized study of combined maternal and infant food allergen avoidance. *J Allergy Clin Immunol* 95: 1179–90.

ZEIGER R S, HELLER S, MELLON M H *et al.* (1989) Effect of combined maternal and infant food-allergen avoidance on development of atopy in early infancy: a randomised study. *J Allergy Clin Immunol* 84: 72–89.

ZEIGER R S, SAMPSON H A, BOCK S A *et al.* (1999) Soy allergy in infants and children with IgE-associated cows' milk allergy. *J Pediatr* 134: 614–22.

Index